ALGEBRA Book 5
LINEAR GRAPHS

Contents

Contact the Author

My name is Grace Hartford, the author of this book.

You can email me at **grace@smile.ws**

I have spent countless hours on this book
to create the best experience for my readers.

I would love to hear any feedback. Alternatively, if you have any issues,
please email me, and I will sort them out.

Copyright

Other Math Books

Arithmetic Series

Practice arithmetic problems
Visit: **smile.ws/pma9**

Multi-Step Math Series

Multi-step math problems
Visit: **smile.ws/pmm9**

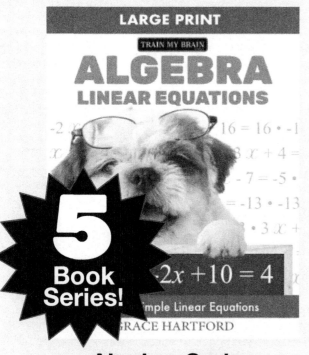

Algebra Series

5 types of problems
Visit: **smile.ws/pmx9**

Turn to pages 76–79 for more puzzle books and bonus samples!

Join our free club for discount coupons on future books!

Visit: **smile.ws/cmx9**

Welcome

This book of algebra problems offers a fun way to train your brain by keeping your mind active! To provide the largest number of problems, this book contains no tuition. But you can email me at grace@smile.ws for a list of online resources that provide guidance.

Extra Wide Margins – Rip out the Pages!

Every page has extra-wide margins. So you can easily rip out the pages, which can make it more convenient to solve the problems.

More Algebra Books Available

There are five books available in this series:

Book 1: Simple Linear Equations

Book 2: Hard Linear Equations

Book 3: Quadratic Factoring

Book 4: Hard Quadratic Factoring

Book 5: Linear Graphs

Collect them all for maximum challenge! (This book is **Book 5**)

Enjoy the book. Love,

Grace xxx

Grace Hartford – Founder of TRAIN MY BRAIN Global

Turn to pages 76-79 for bonus samples of my other math books!

Collect all 5 books in this Algebra series!

Visit: **smile.ws/pmx9**

Linear Graphs

Give the equation for each line in the form y = m x + b

1)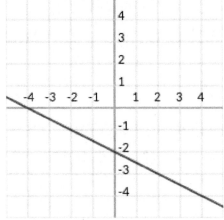

2)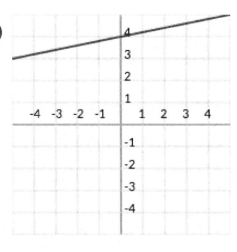

3)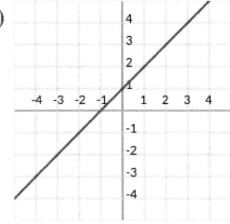

4)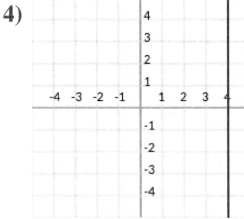

5)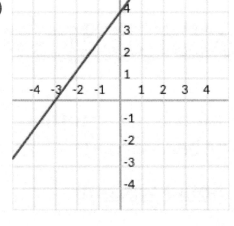

6)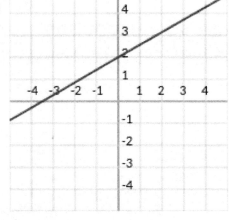

Linear Graphs

Give the equation for each line in the form y = m x + b

...Rip out the pages!

7)

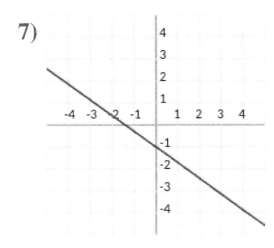

8)

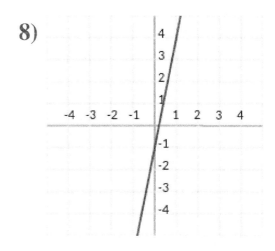

9)

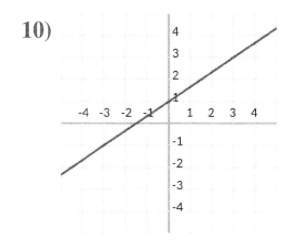

10)

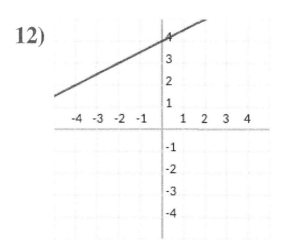

11)

12)

Linear Graphs

Give the equation for each line in the form y = m x + b

13)

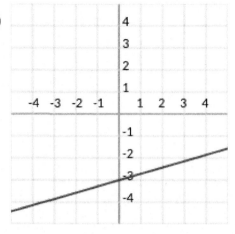

14)

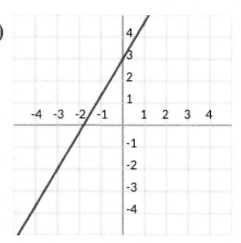

15)

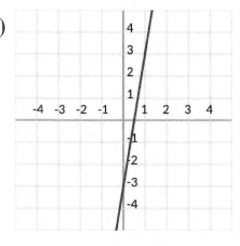

16)

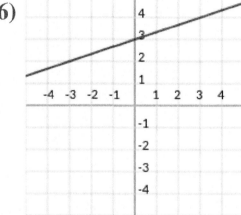

17)

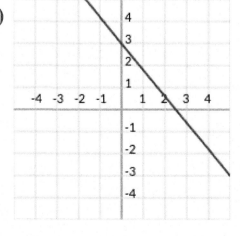

18)

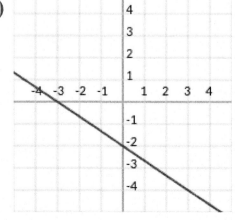

Linear Graphs

Give the equation for each line in the form y = m x + b

19)

20)

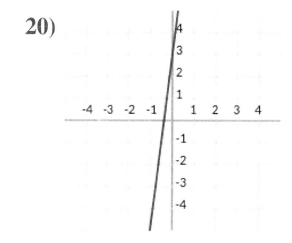

21)

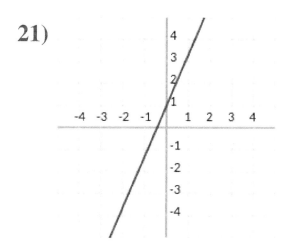

22)

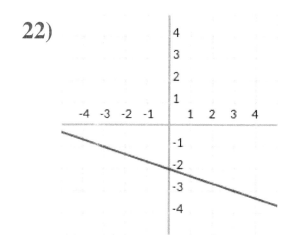

23)

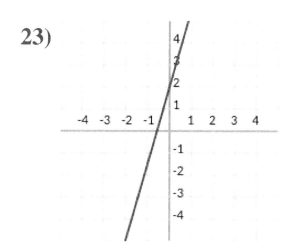

24)

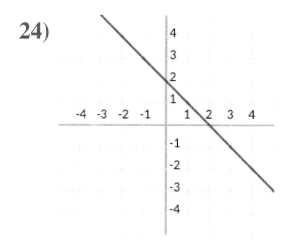

Linear Graphs

Give the equation for each line in the form y = m x + b

25)

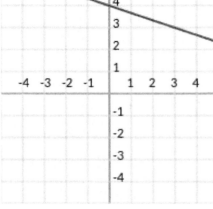

26)

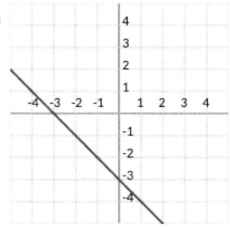

27)

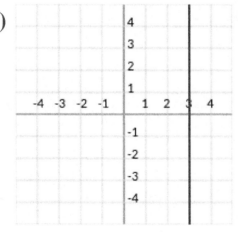

28)

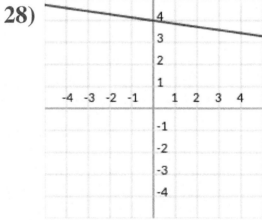

29)

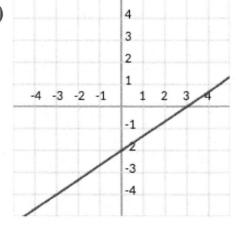

30)

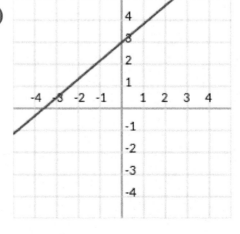

Linear Graphs

Give the equation for each line in the form y = m x + b

31)

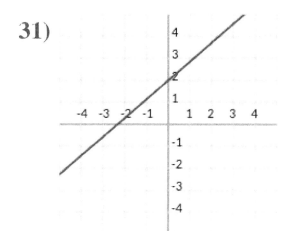

32)

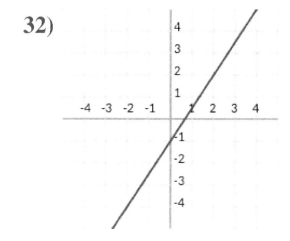

33)

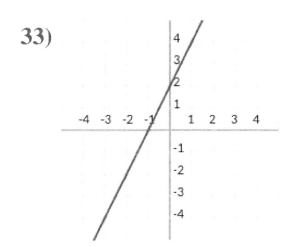

34)

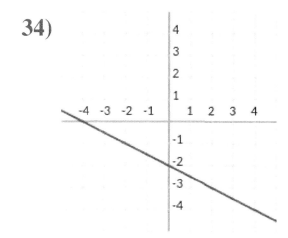

35)

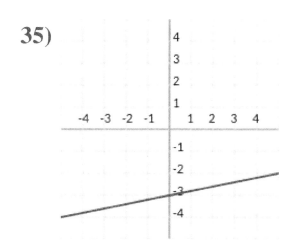

36)

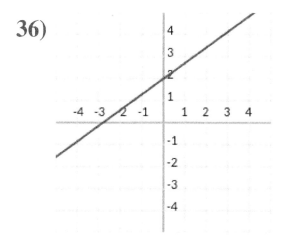

Linear Graphs

Give the equation for each line in the form y = m x + b

37)

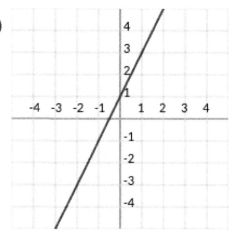

38)

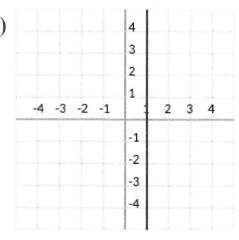

39)

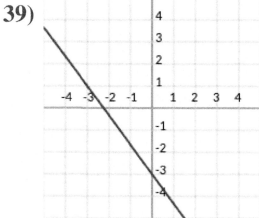

40)

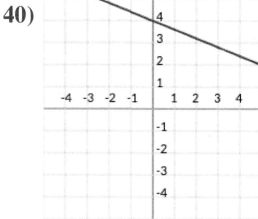

41)

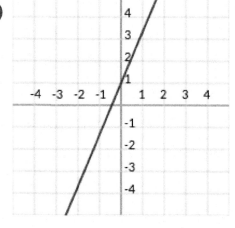

42)

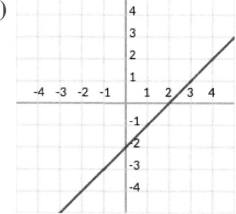

Linear Graphs

Give the equation for each line in the form y = m x + b

43)

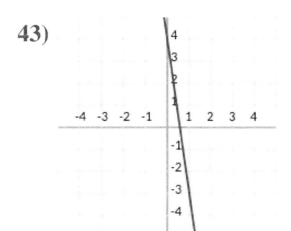

44)

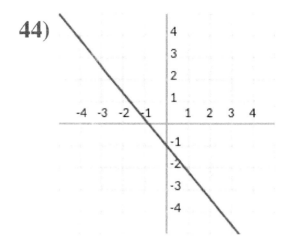

45)

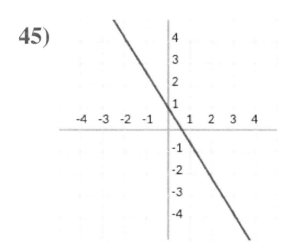

46)

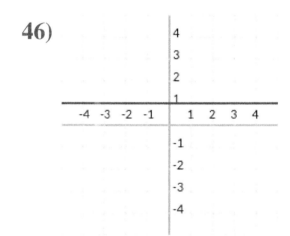

47)

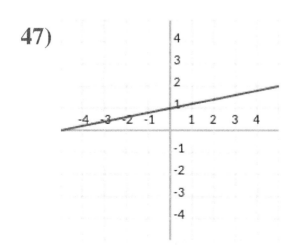

48)

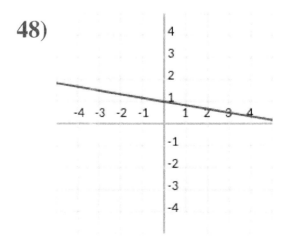

Linear Graphs

Give the equation for each line in the form y = m x + b

49)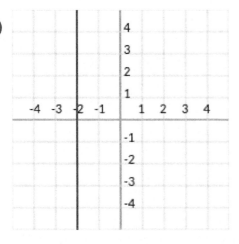

50)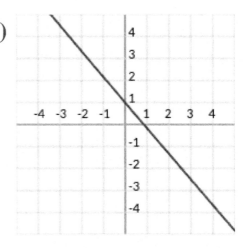

51)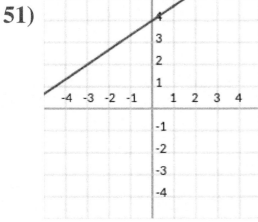

52)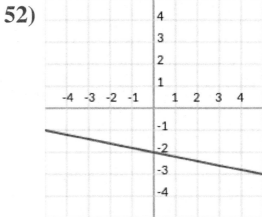

53)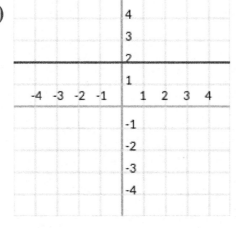

54)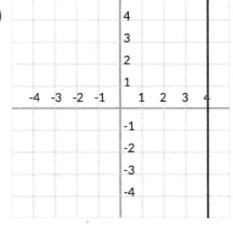

Linear Graphs

Give the equation for each line in the form y = m x + b

55)

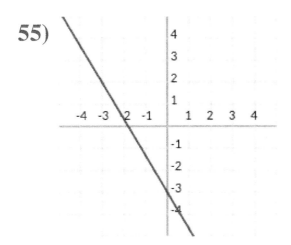

56)

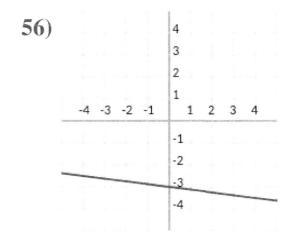

57)

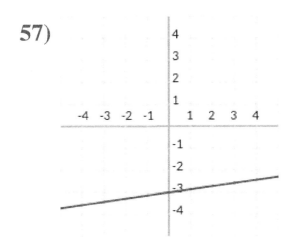

58)

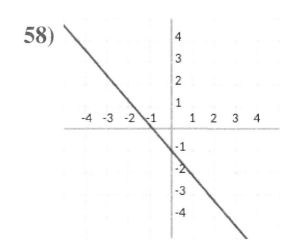

59)

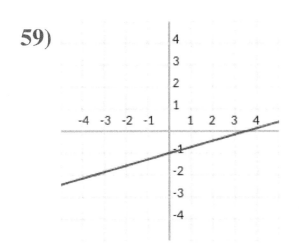

60)

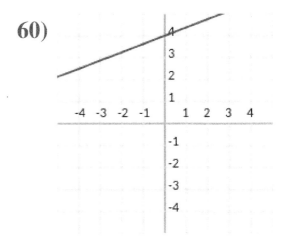

Linear Graphs

Give the equation for each line in the form y = m x + b

61)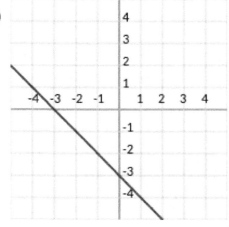

62)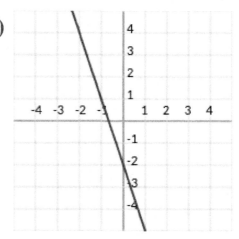

63)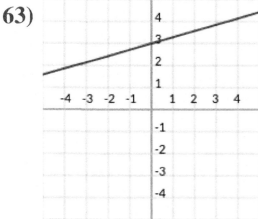

64)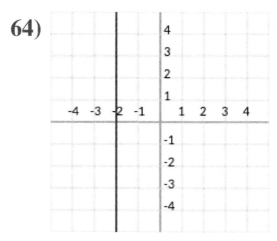

65)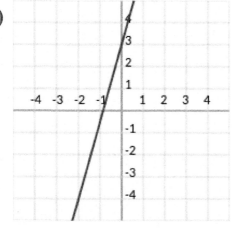

66)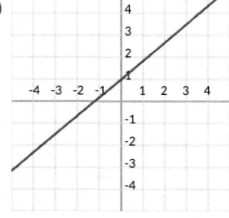

Linear Graphs

Give the equation for each line in the form y = m x + b

67)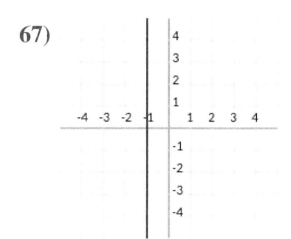

68)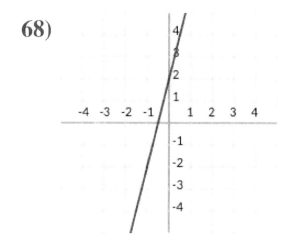

69)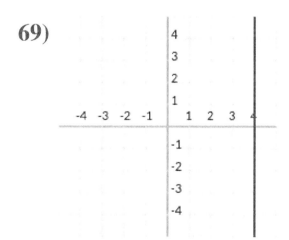

70)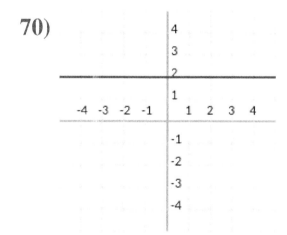

71)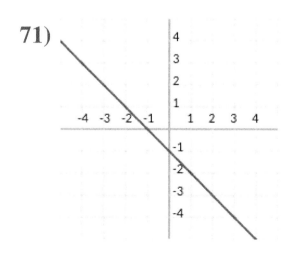

72)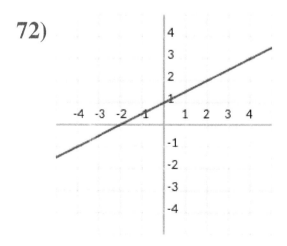

Linear Graphs

Give the equation for each line in the form y = m x + b

73)

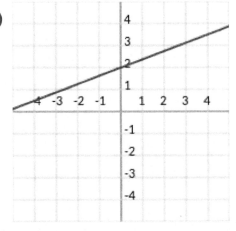

74)

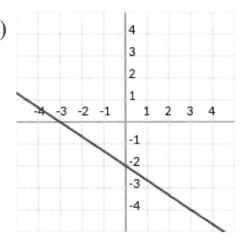

75)

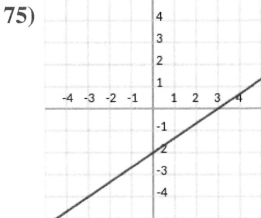

76)

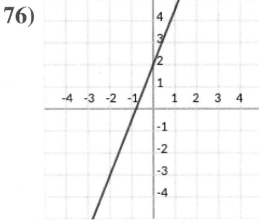

77)

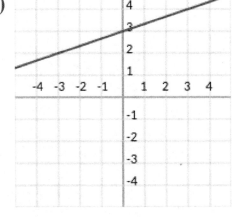

78)

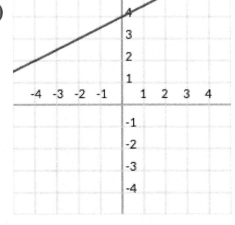

Linear Graphs

Give the equation for each line in the form y = m x + b

79)

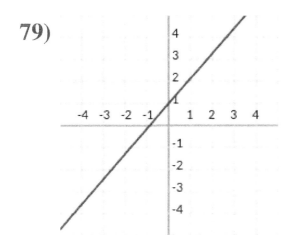

80)

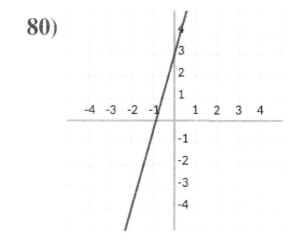

81)

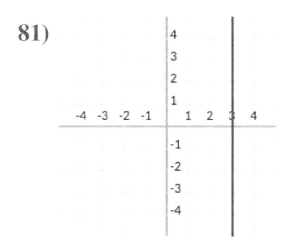

82)

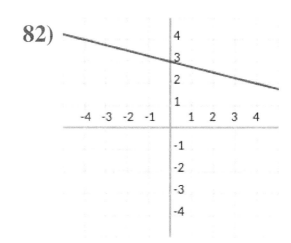

83)

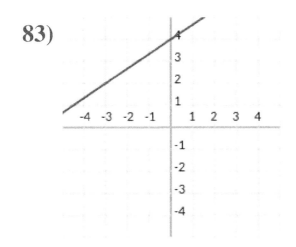

84)

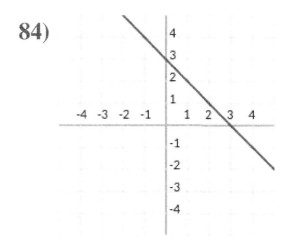

Linear Graphs

Give the equation for each line in the form y = m x + b

85)
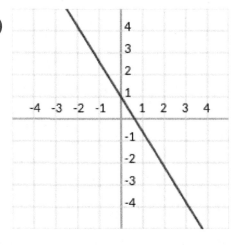

86)
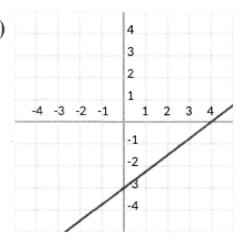

87)
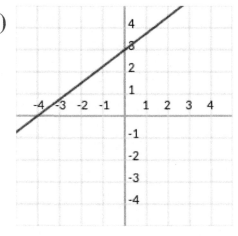

88)
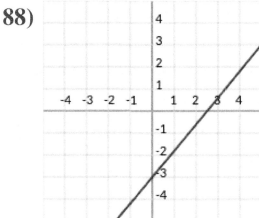

89)
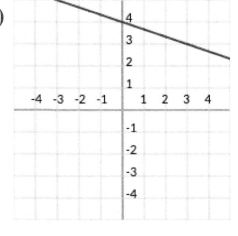

90)
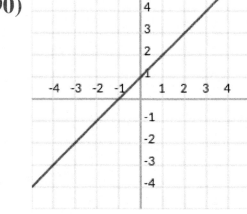

Linear Graphs

Give the equation for each line in the form y = m x + b

91)

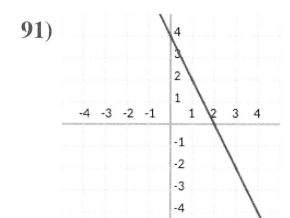

92)

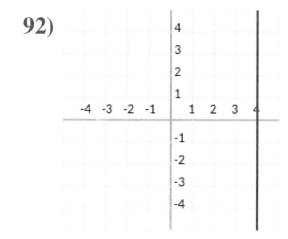

93)

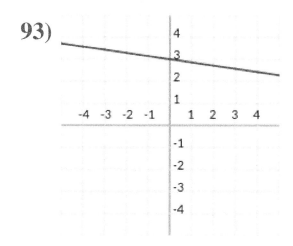

94)

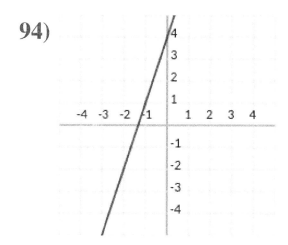

95)

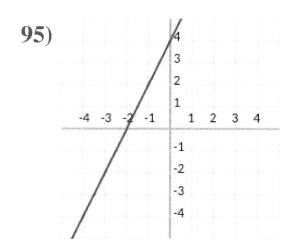

96)

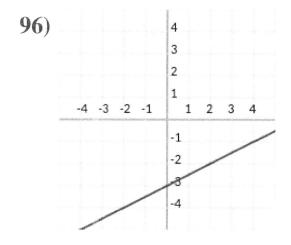

Linear Graphs

Give the equation for each line in the form y = m x + b

97)

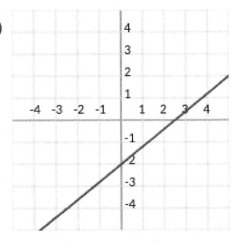

98)

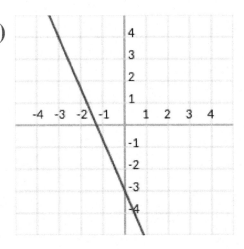

99)

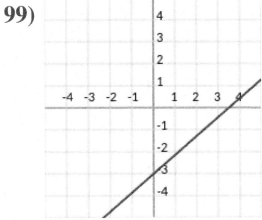

100)

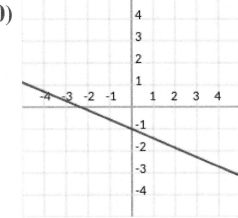

101)

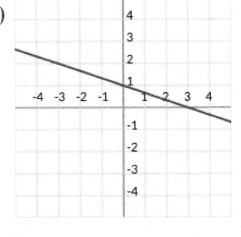

102)

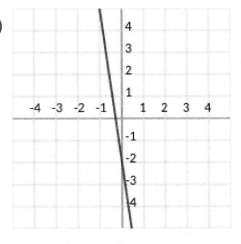

Linear Graphs

Give the equation for each line in the form y = m x + b

103)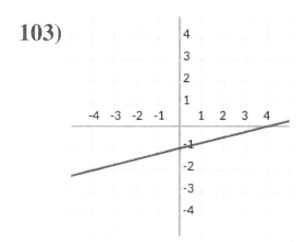

104)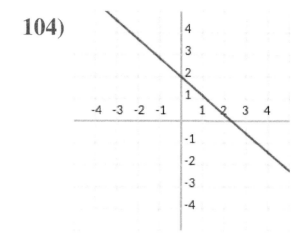

105)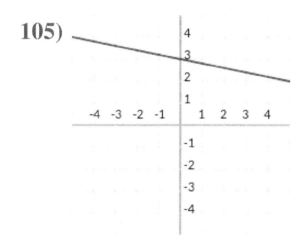

106)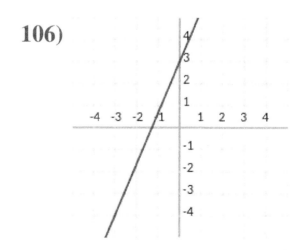

107)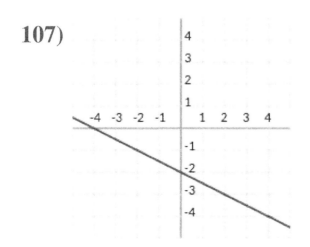

108)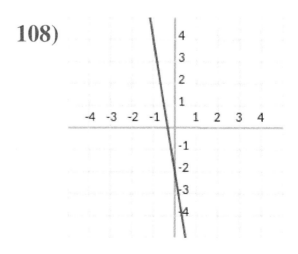

Linear Graphs

Give the equation for each line in the form y = m x + b

109)

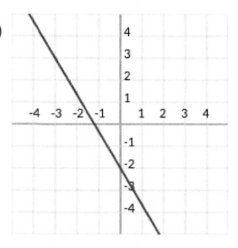

110)

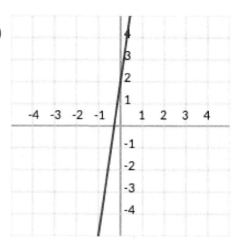

111)

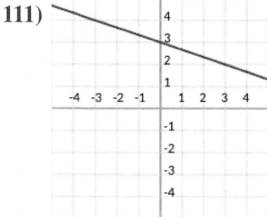

112)

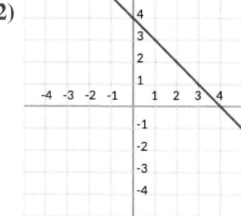

113)

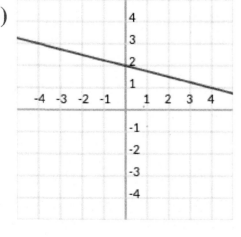

114)

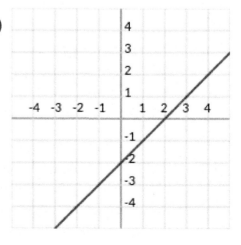

Linear Graphs

Give the equation for each line in the form y = m x + b

115)

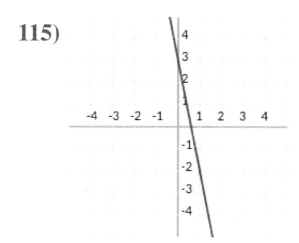

116)

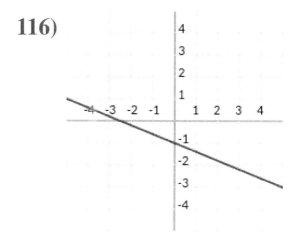

117)

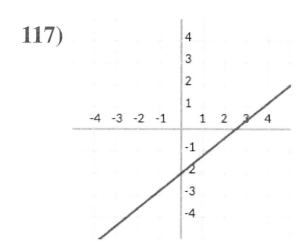

118)

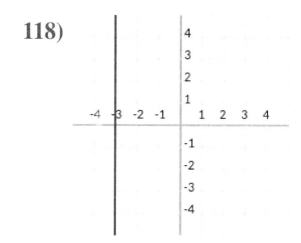

119)

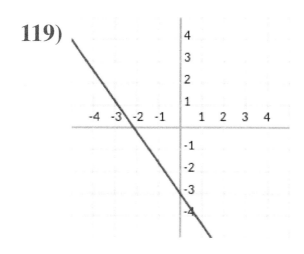

120)

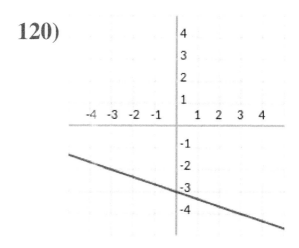

Linear Graphs

Give the equation for each line in the form y = m x + b

121)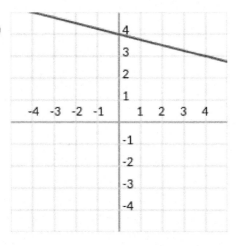

122)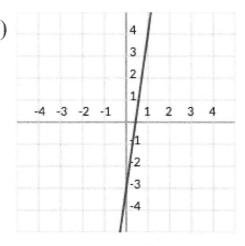

123)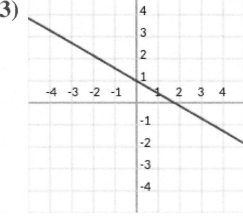

124)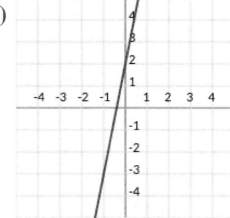

125)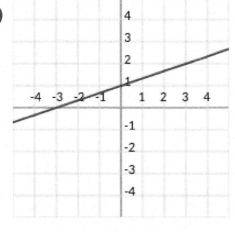

126)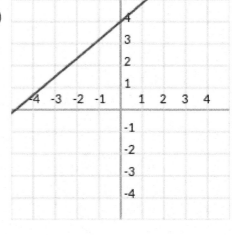

Linear Graphs

Give the equation for each line in the form y = m x + b

127)

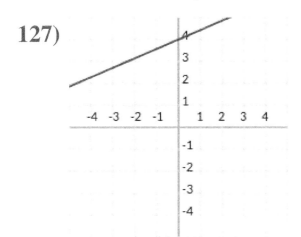

128)

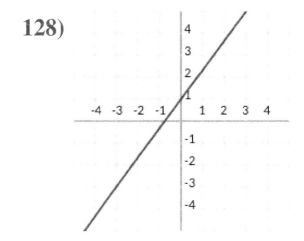

129)

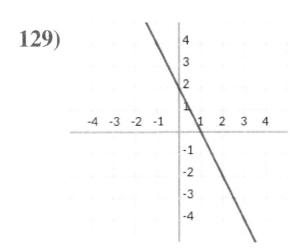

130)

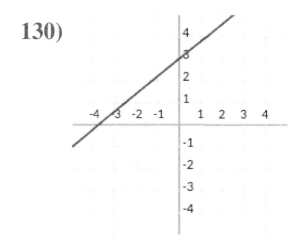

131)

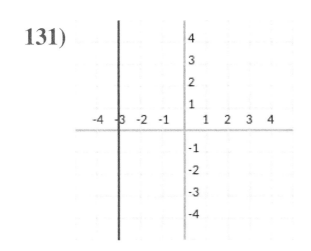

132)

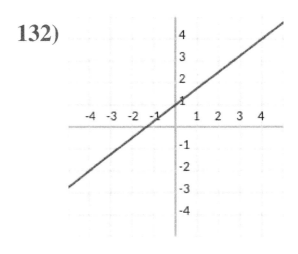

Linear Graphs

Give the equation for each line in the form y = m x + b

133)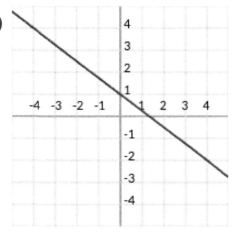

134)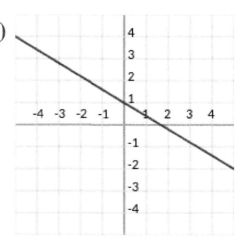

135)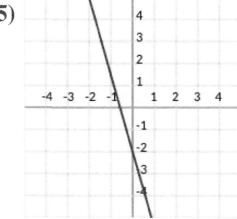

136)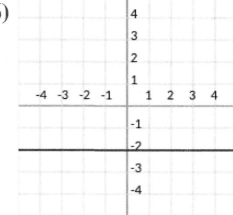

137)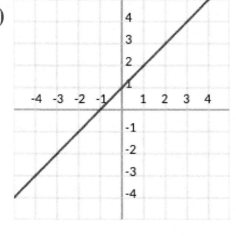

138)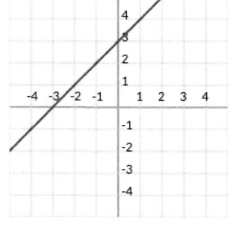

Linear Graphs

Give the equation for each line in the form y = m x + b

139)

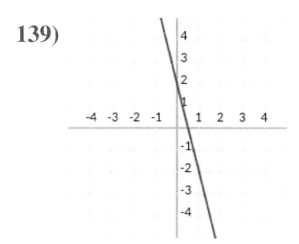

140)

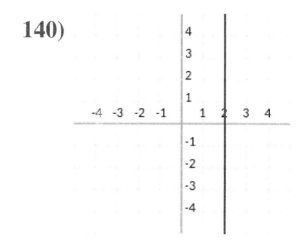

141)

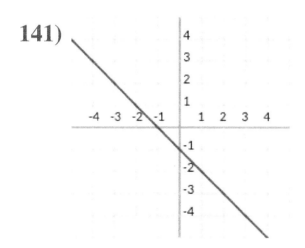

142)

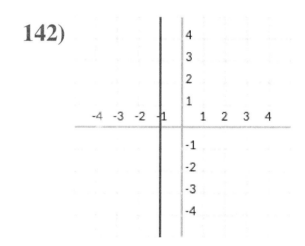

143)

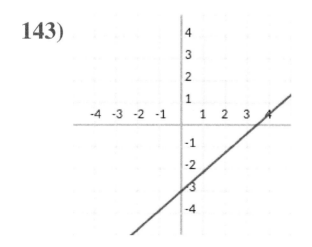

144)

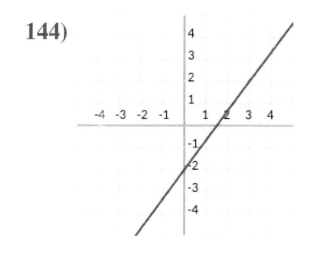

Linear Graphs

Give the equation for each line in the form y = m x + b

145)

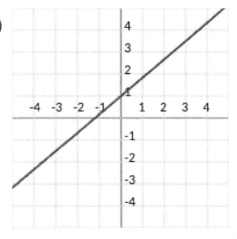

146)

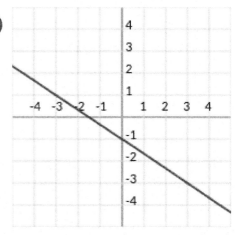

147)

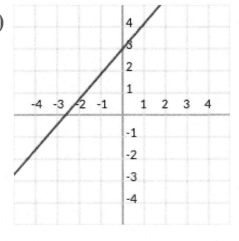

148)

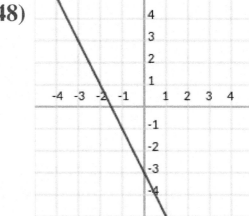

149)

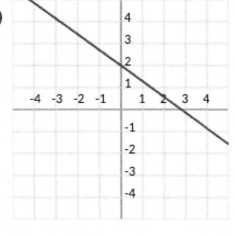

150)

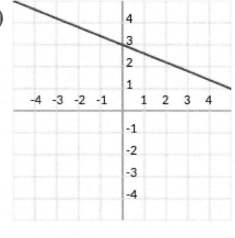

Linear Graphs

Give the equation for each line in the form y = m x + b

151)

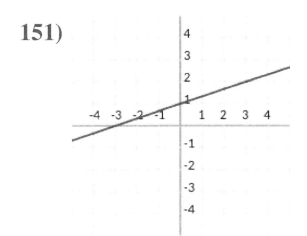

152)

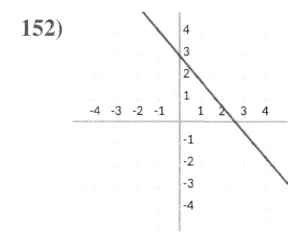

153)

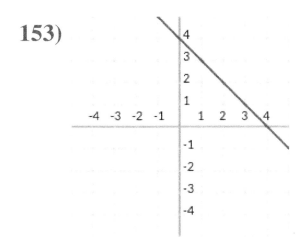

154)

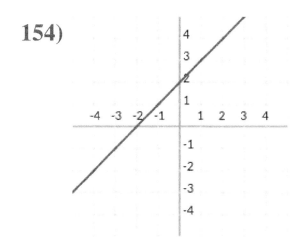

155)

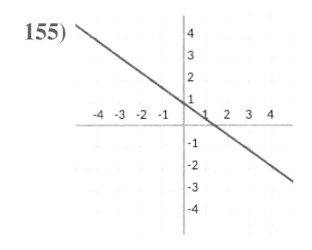

156)

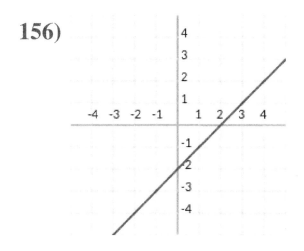

Linear Graphs

Give the equation for each line in the form y = m x + b

157)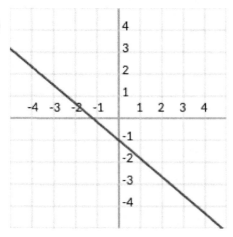

158)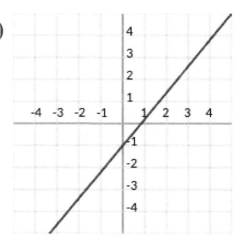

159)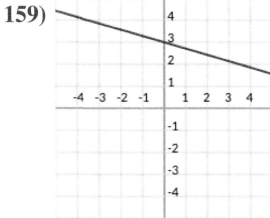

160)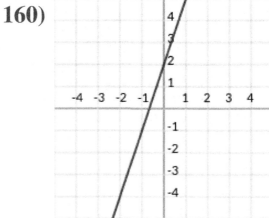

161)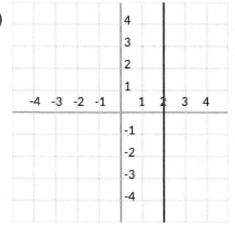

162)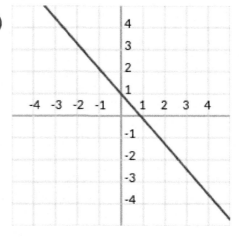

Linear Graphs

Give the equation for each line in the form y = m x + b

163)

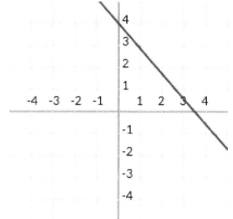

164)

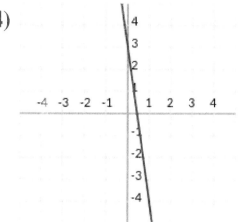

165)

166)

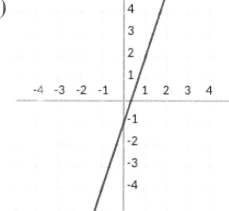

167)

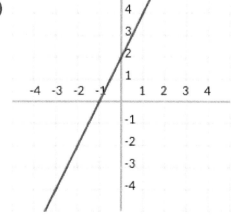

168)

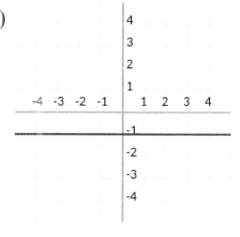

Linear Graphs

Give the equation for each line in the form y = m x + b

169)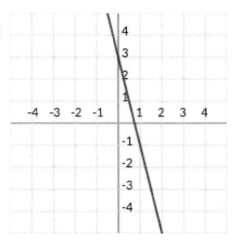

170)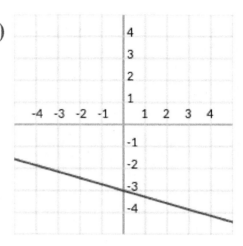

171)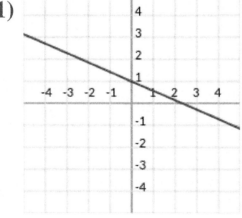

172)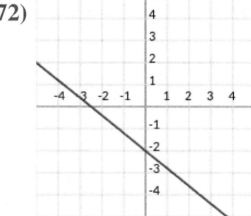

173)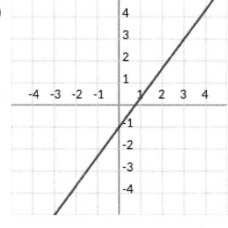

174)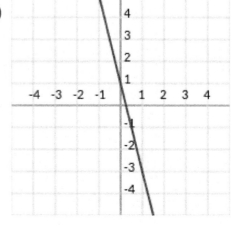

Linear Graphs

Give the equation for each line in the form y = m x + b

175)

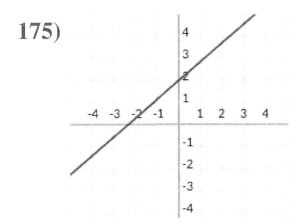

176)

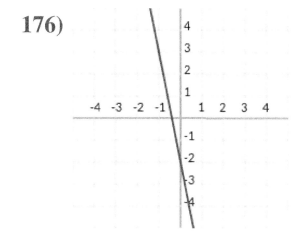

177)

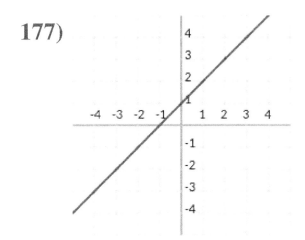

178)

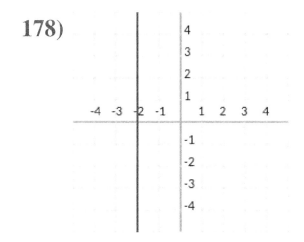

179)

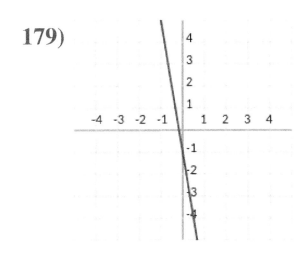

180)

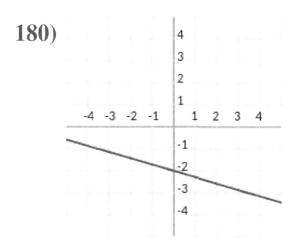

Linear Graphs

Give the equation for each line in the form y = m x + b

181)

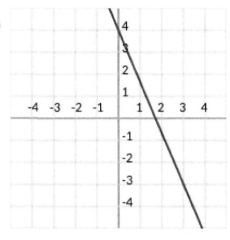

182)

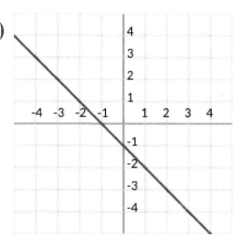

183)

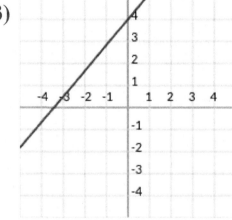

184)

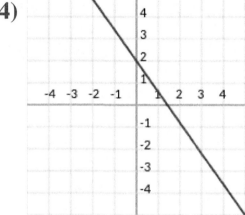

185)

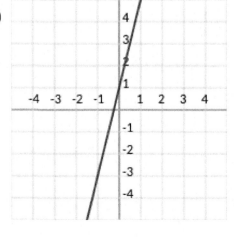

186)

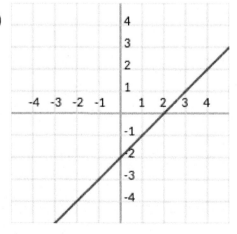

Linear Graphs

Give the equation for each line in the form y = m x + b

187)

188)

189)

190)

191)

192)

Linear Graphs

Give the equation for each line in the form y = m x + b

193)

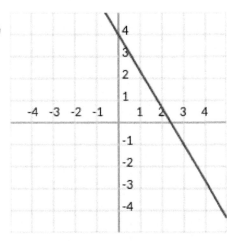

194)

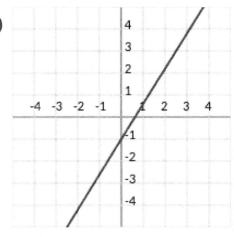

195)

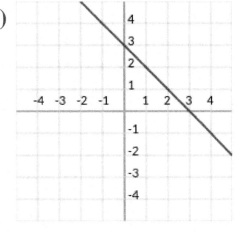

196)

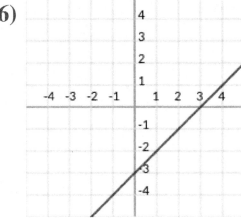

197)

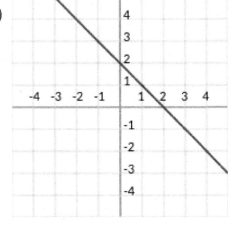

198)

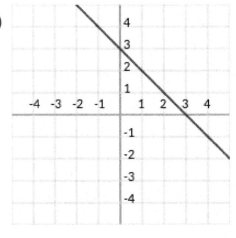

Linear Graphs

Give the equation for each line in the form y = m x + b

199)

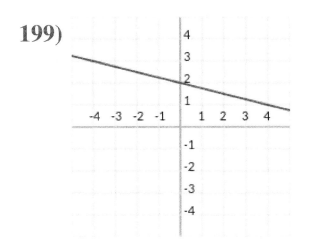

200)

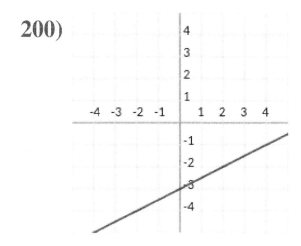

201)

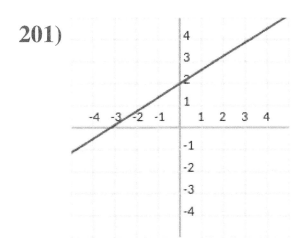

202)

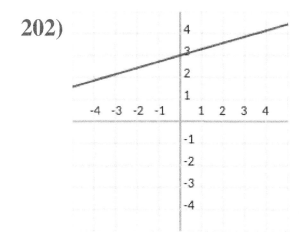

203)

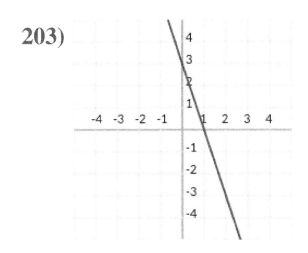

204)

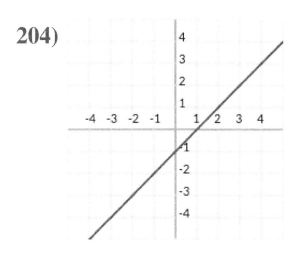

Linear Graphs

Give the equation for each line in the form y = m x + b

205)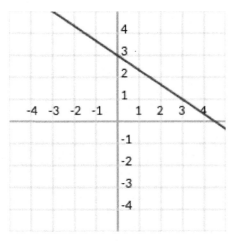

206)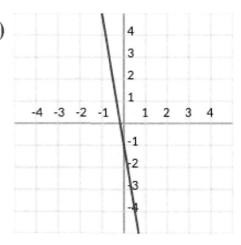

207)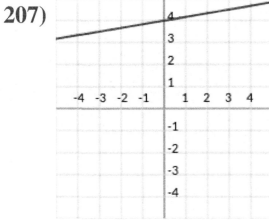

208)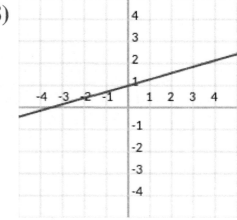

209)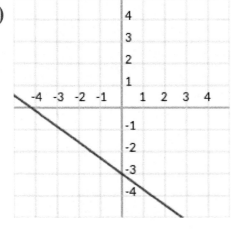

210)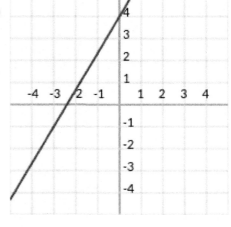

Linear Graphs

Give the equation for each line in the form y = m x + b

211)

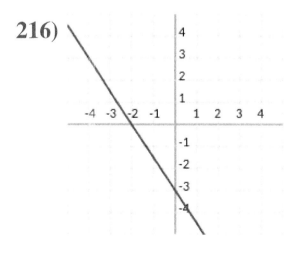

212)

213)

214)

215)

216)

Linear Graphs

Give the equation for each line in the form y = m x + b

217)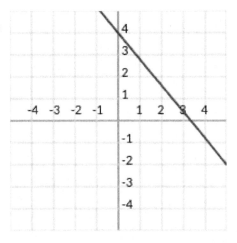

218)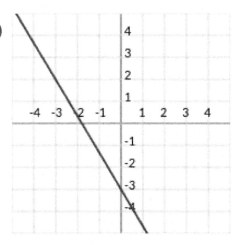

219)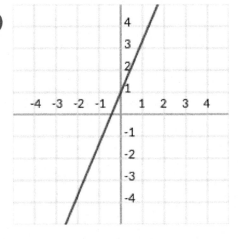

220)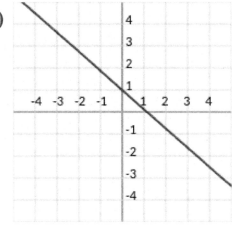

221)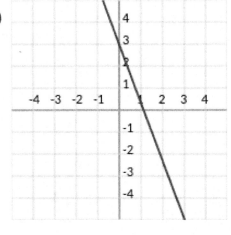

222)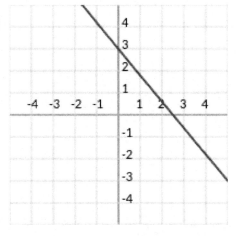

Linear Graphs

Give the equation for each line in the form y = m x + b

223)

224)

225)

226)

227)

228)

Linear Graphs

Give the equation for each line in the form y = m x + b

229)
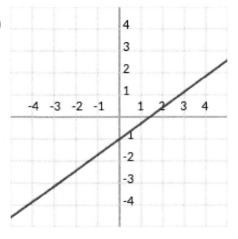

230)
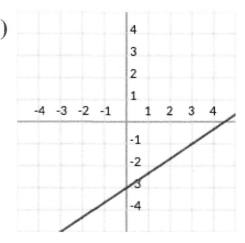

231)
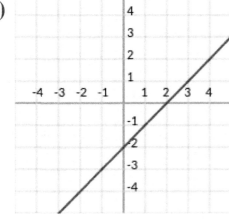

232)
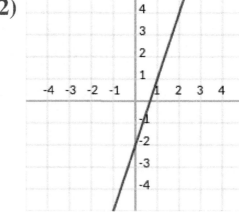

233)
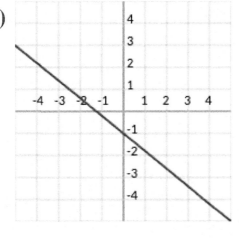

234)
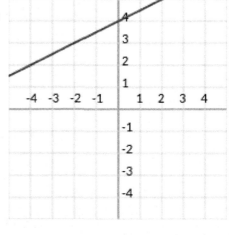

Linear Graphs

Give the equation for each line in the form y = m x + b

235)

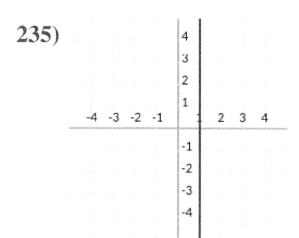

236)

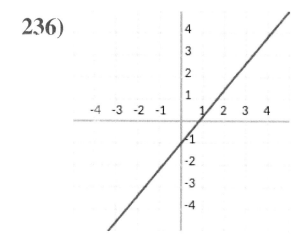

237)

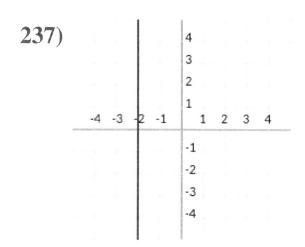

238)

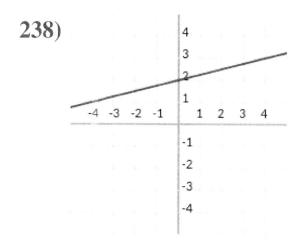

239)

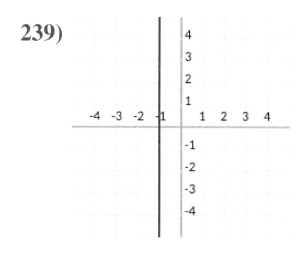

240)

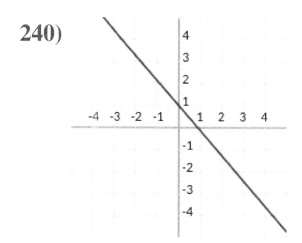

43

Linear Graphs

Give the equation for each line in the form y = m x + b

241)

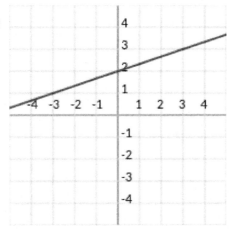

242)

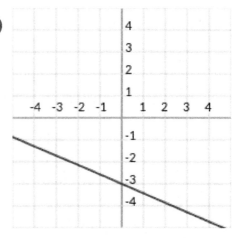

243)

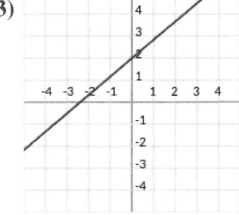

244)

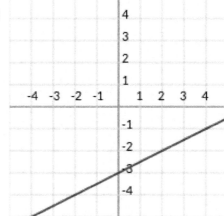

245)

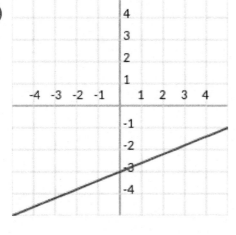

246)

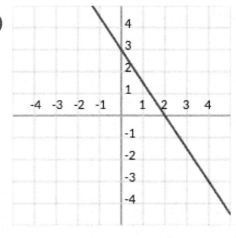

Linear Graphs

Give the equation for each line in the form y = m x + b

247)

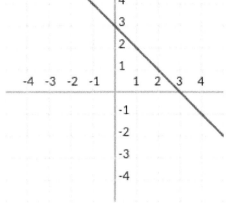

248)

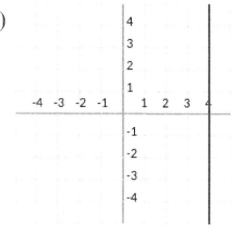

249)

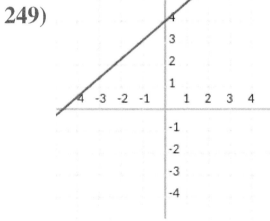

250)

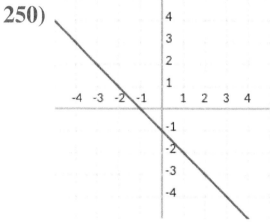

251)

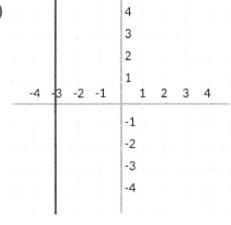

252)

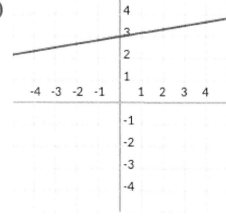

Linear Graphs

Give the equation for each line in the form y = m x + b

253)

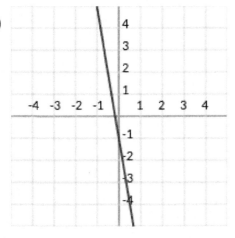

254)

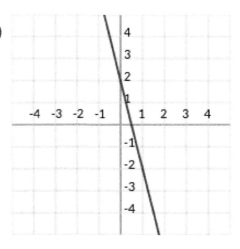

255)

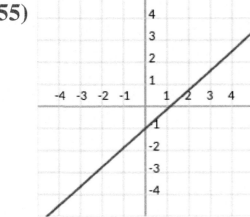

256)

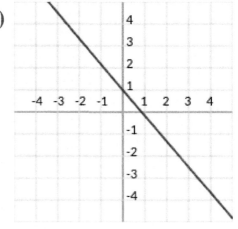

257)

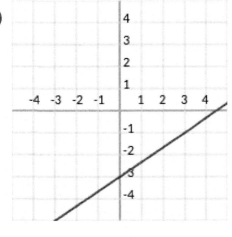

258)

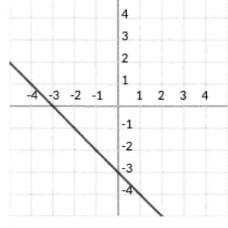

Linear Graphs

Give the equation for each line in the form y = m x + b

259)

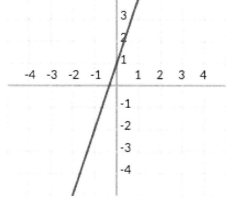

260)

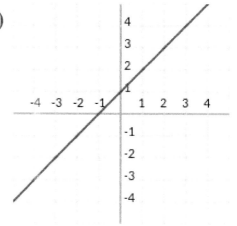

261)

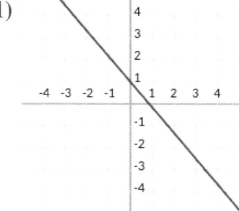

262)

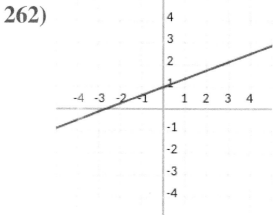

263)

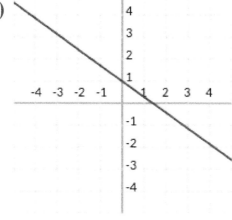

264)

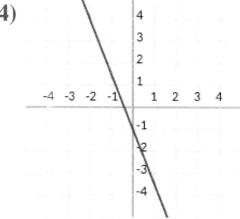

Linear Graphs

Give the equation for each line in the form y = m x + b

265)

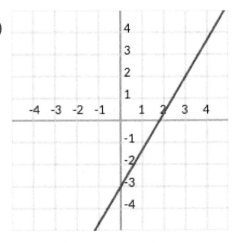

266)

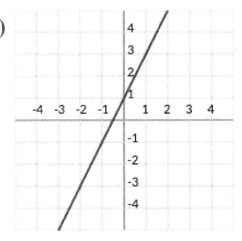

267)

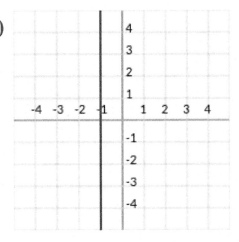

268)

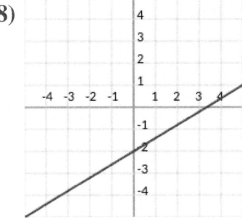

269)

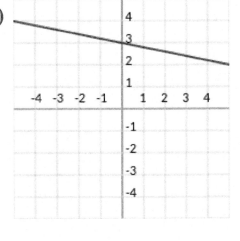

270)

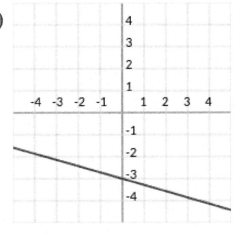

Linear Graphs

Give the equation for each line in the form y = m x + b

271)

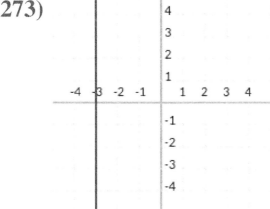

272)

273)

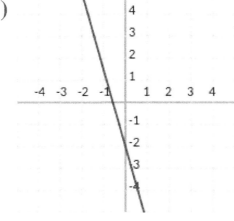

274)

275)

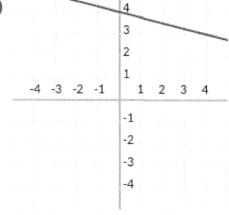

276)

Linear Graphs

Give the equation for each line in the form y = m x + b

277)

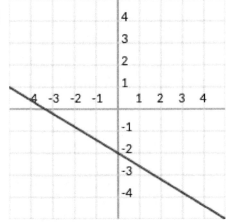

278)

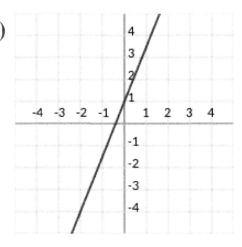

279)

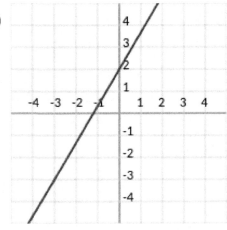

280)

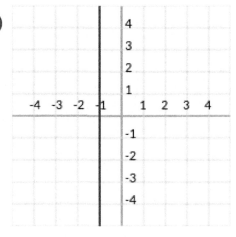

281)

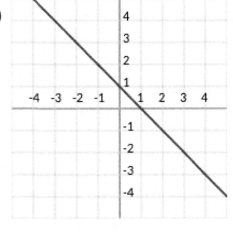

282)

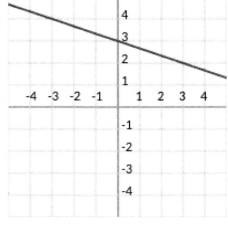

Linear Graphs

Give the equation for each line in the form y = m x + b

283)

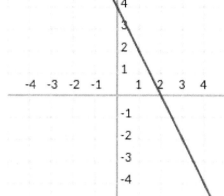

284)

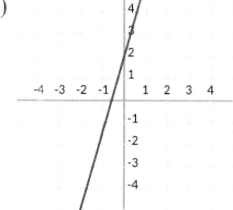

285)

286)

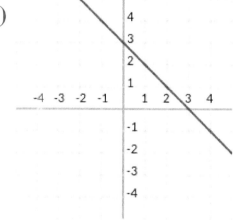

287)

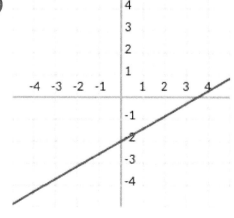

288)

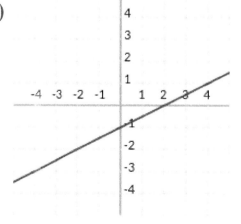

Linear Graphs

Give the equation for each line in the form y = m x + b

289)

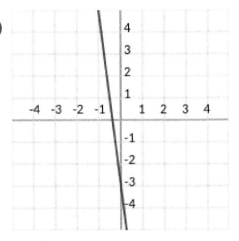

290)

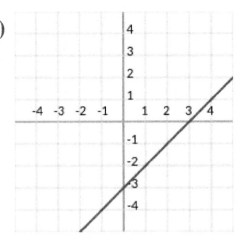

291)

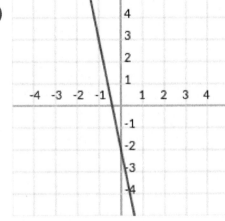

292)

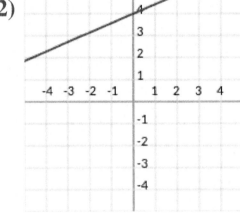

293)

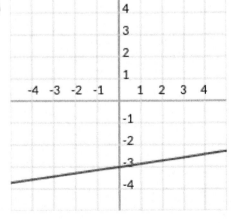

294)

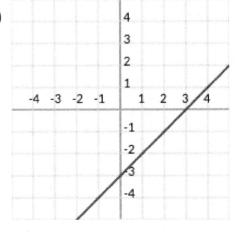

Linear Graphs

Give the equation for each line in the form y = m x + b

295)

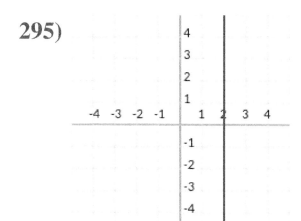

296)

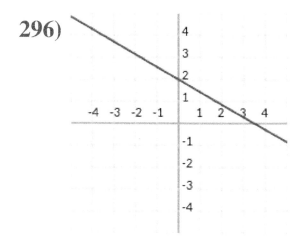

297)

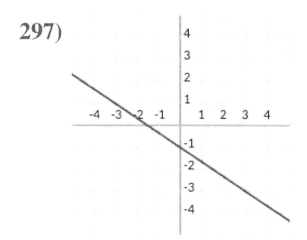

298)

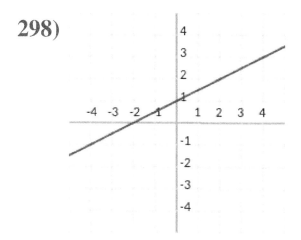

299)

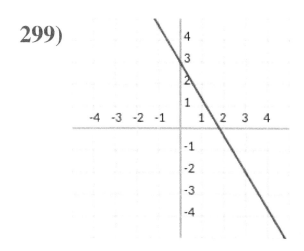

300)

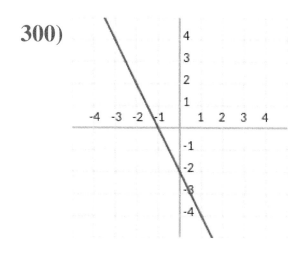

Linear Graphs

Give the equation for each line in the form y = m x + b

301)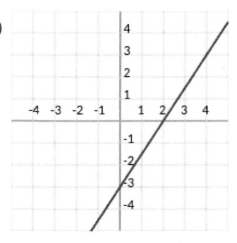

302)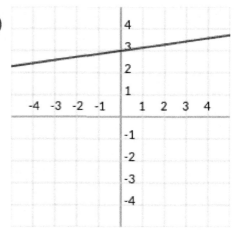

303)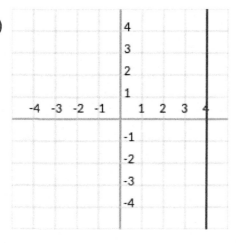

304)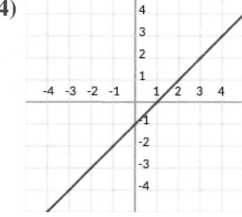

305)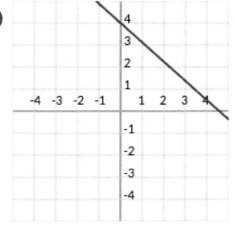

306)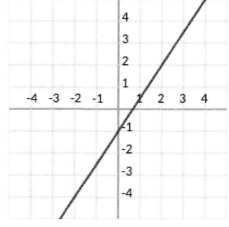

Linear Graphs

Give the equation for each line in the form y = m x + b

307)

308)

309)

310)

311)

312)

Linear Graphs

Give the equation for each line in the form y = m x + b

313)

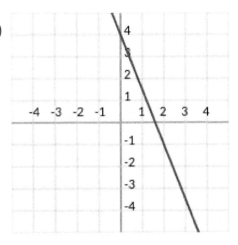

314)

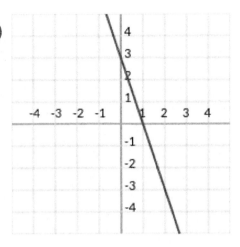

315)

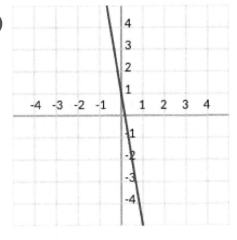

316)

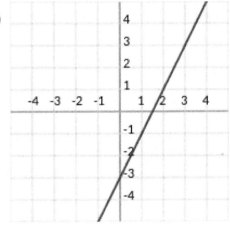

317)

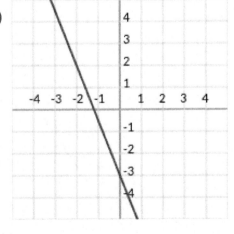

318)

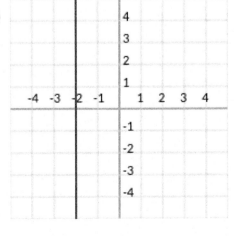

Linear Graphs

Give the equation for each line in the form y = m x + b

319)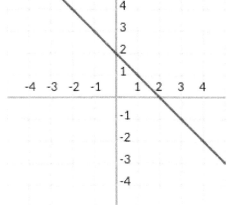

320)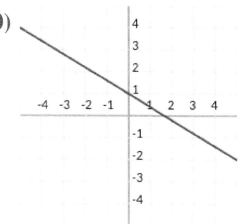

321)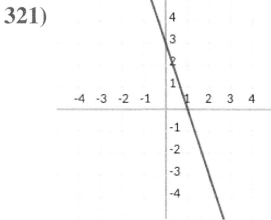

322)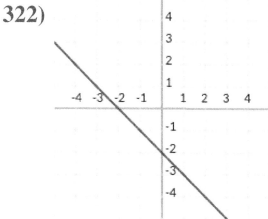

323)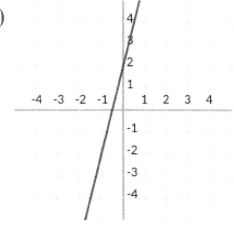

324)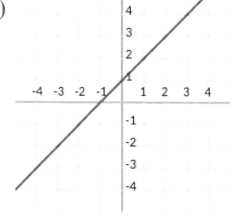

Linear Graphs

Give the equation for each line in the form y = m x + b

325)

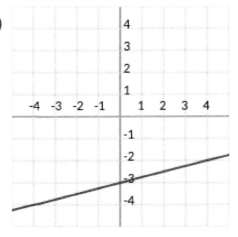

326)

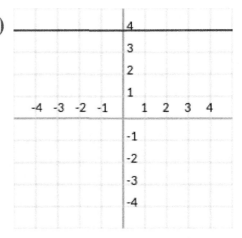

327)

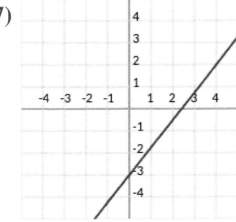

328)

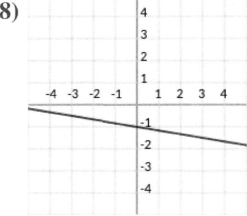

329)

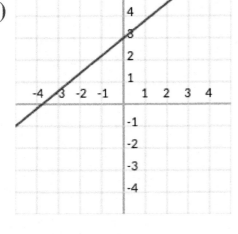

330)

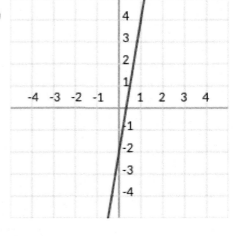

Linear Graphs

Give the equation for each line in the form y = m x + b

331)

332)

333)

334)

335)

336)

Linear Graphs

Give the equation for each line in the form y = m x + b

337)

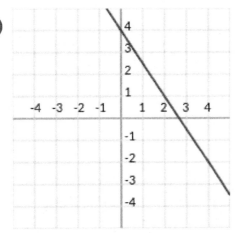

338)

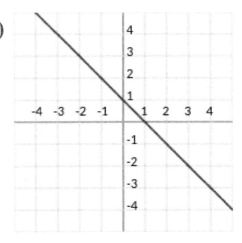

339)

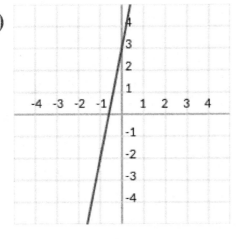

340)

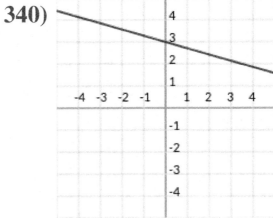

341)

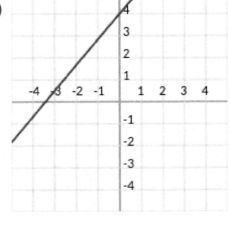

342)

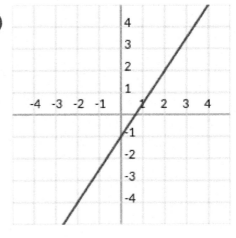

Linear Graphs

Give the equation for each line in the form y = m x + b

343)

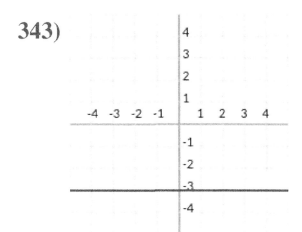

344)

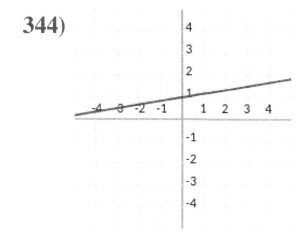

345)

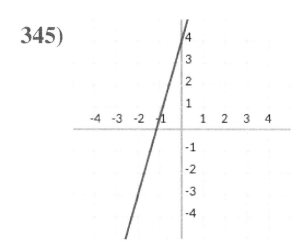

346)

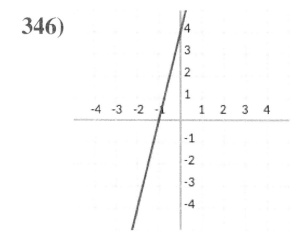

347)

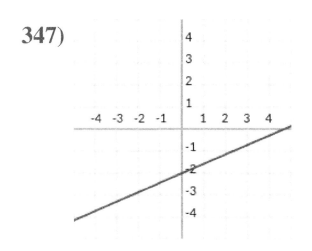

348)

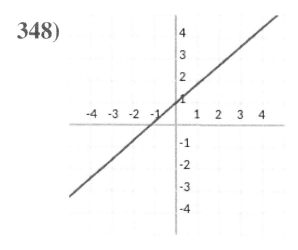

Linear Graphs

Give the equation for each line in the form y = m x + b

349)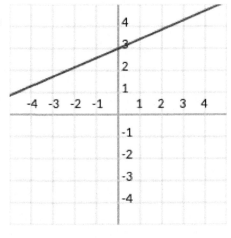

350)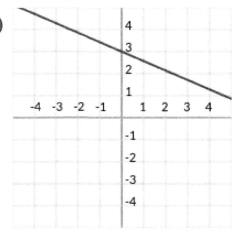

351)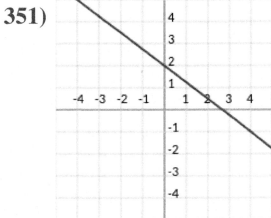

352)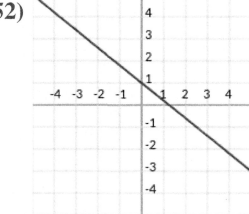

353)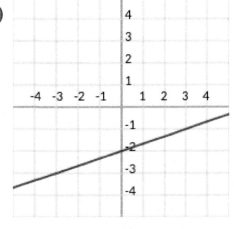

354)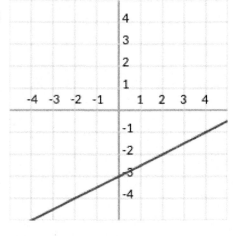

Linear Graphs

Give the equation for each line in the form y = m x + b

355)

356)

357)

358)

359)

360)

Linear Graphs

Give the equation for each line in the form y = m x + b

361)

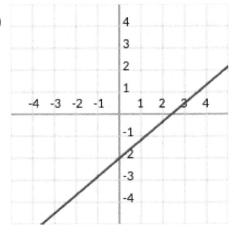

362)

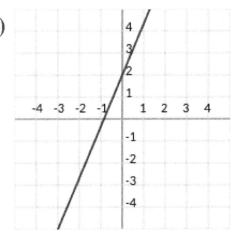

363)

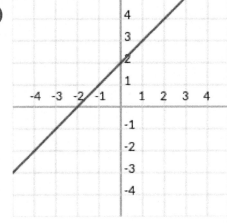

364)

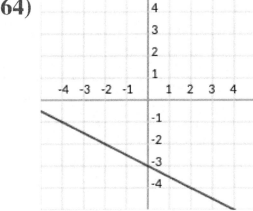

365)

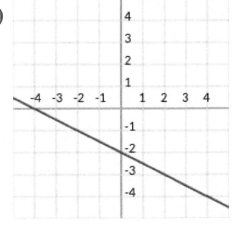

366)

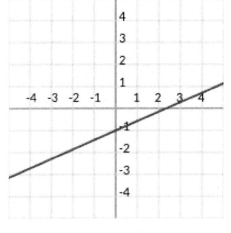

Linear Graphs

Give the equation for each line in the form y = m x + b

367)

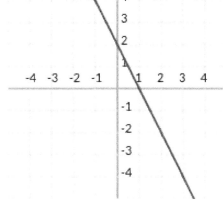

368)

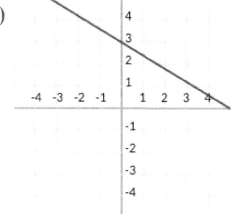

369)

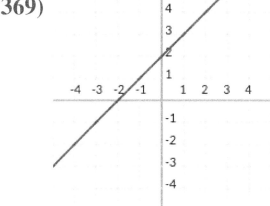

370)

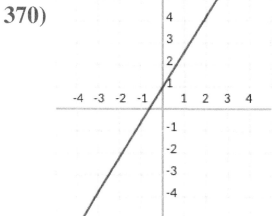

371)

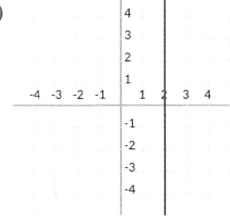

372)

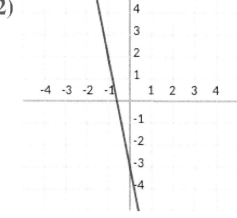

Linear Graphs

Give the equation for each line in the form y = m x + b

373)

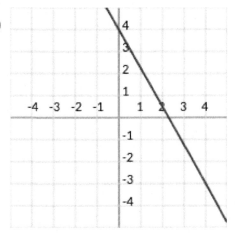

374)

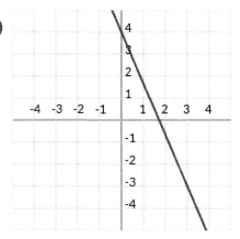

375)

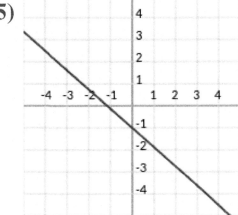

376)

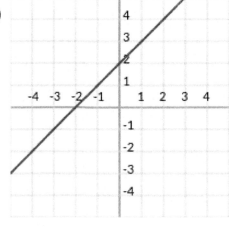

377)

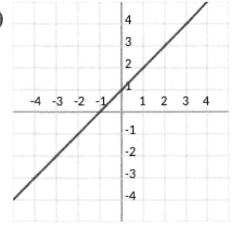

378)

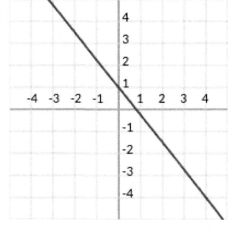

Linear Graphs

Give the equation for each line in the form y = m x + b

379)

380)

381)

382)

383)

384)

Linear Graphs

Give the equation for each line in the form y = m x + b

385)

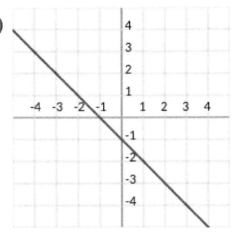

386)

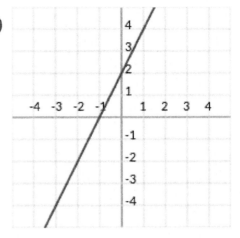

387)

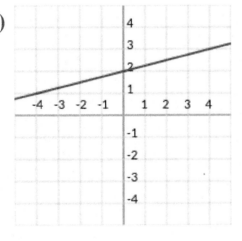

388)

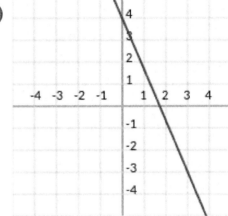

389)

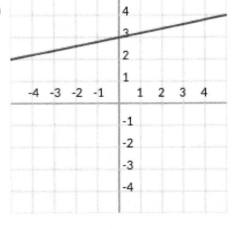

390)

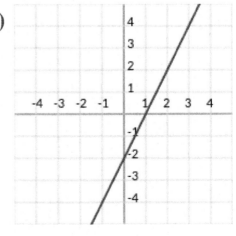

Linear Graphs

Give the equation for each line in the form y = m x + b

391)

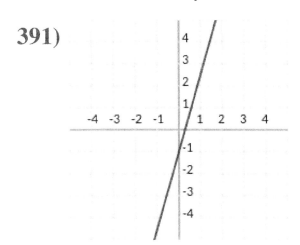

392)

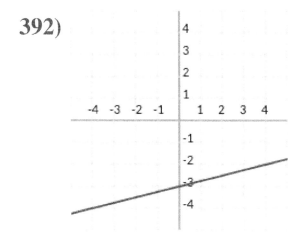

393)

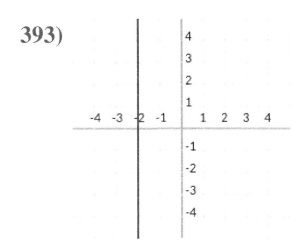

394)

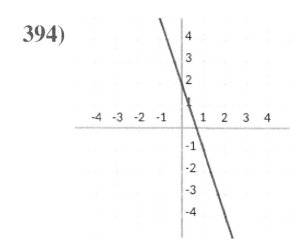

395)

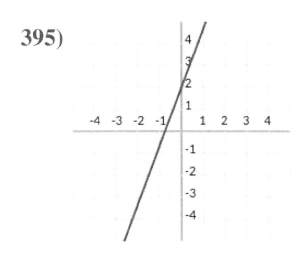

396)

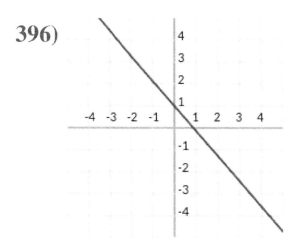

Linear Graphs

Give the equation for each line in the form y = m x + b

397)

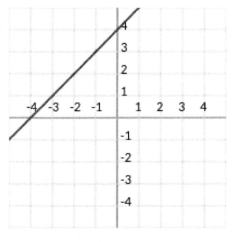

398)

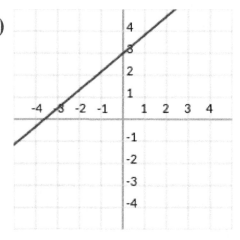

399)

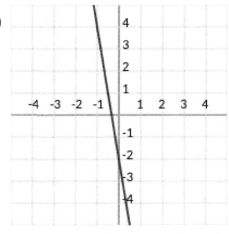

400)

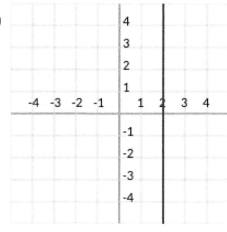

401)

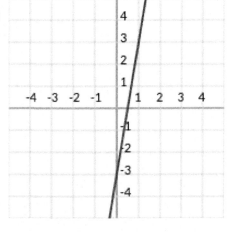

402)

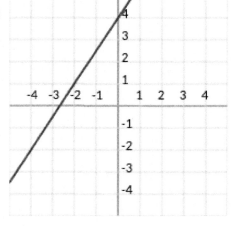

Linear Graphs

Give the equation for each line in the form y = m x + b

403)

404)

405)

406)

407)

408)

Linear Graphs

Give the equation for each line in the form y = m x + b

409)

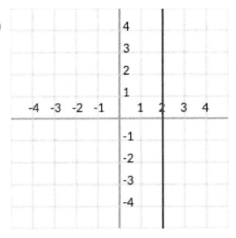

410)

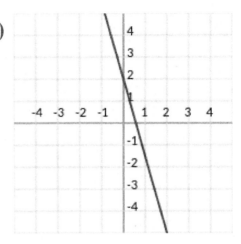

411)

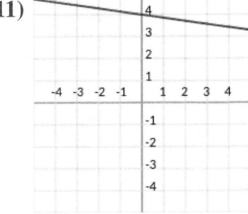

412)

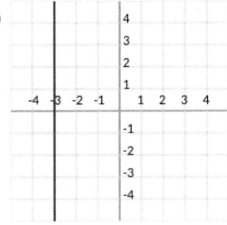

413)

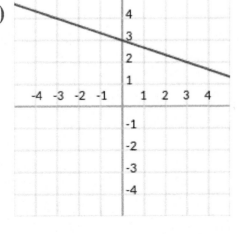

414)

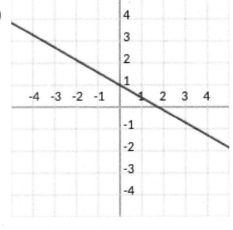

Linear Graphs

Give the equation for each line in the form y = m x + b

415)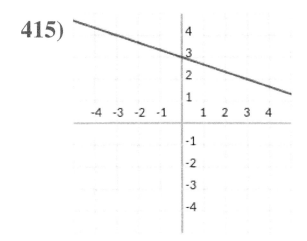

416)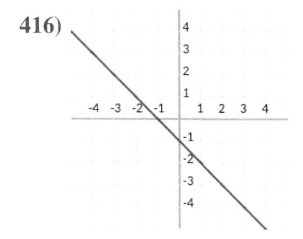

417)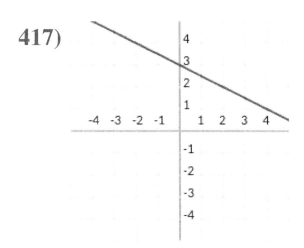

418)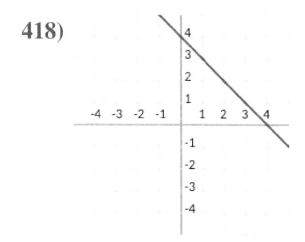

419)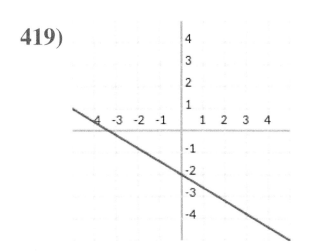

420)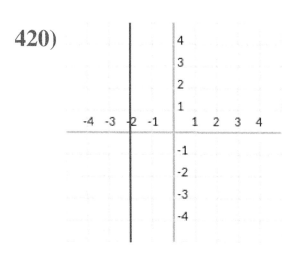

Linear Graphs

Give the equation for each line in the form y = m x + b

421)

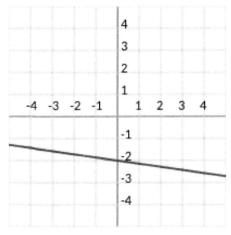

422)

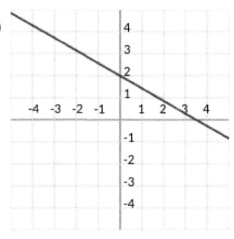

423)

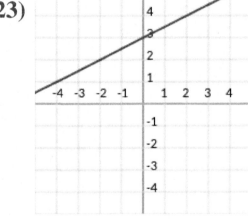

424)

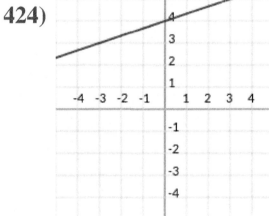

425)

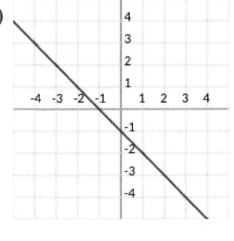

426)

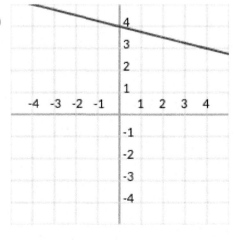

Linear Graphs

Give the equation for each line in the form y = m x + b

427)

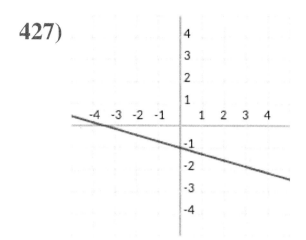

428)

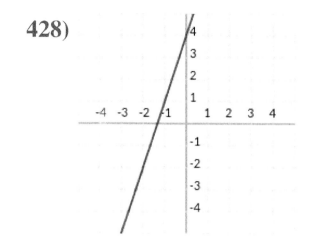

429)

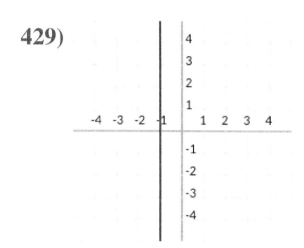

430)

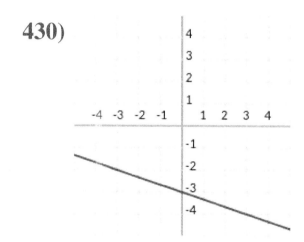

431)

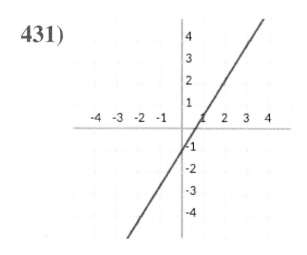

432)

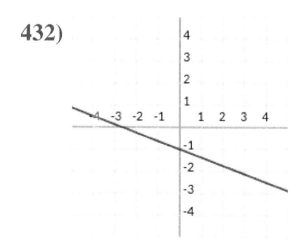

Linear Graphs

Give the equation for each line in the form y = m x + b

433)

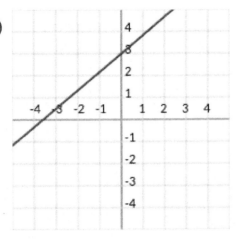

434)

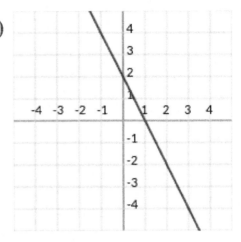

435)

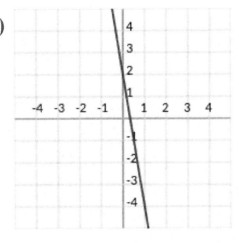

436)

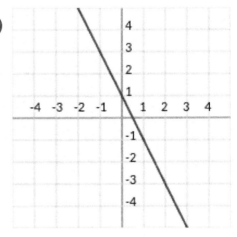

437)

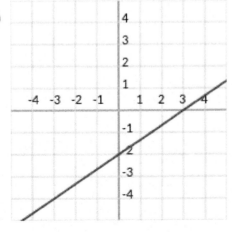

438)

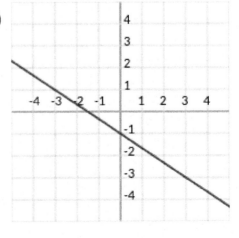

Linear Graphs

Give the equation for each line in the form y = m x + b

439)

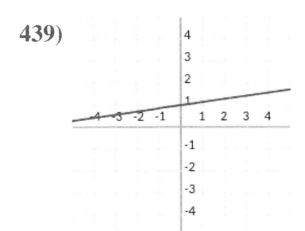

440)

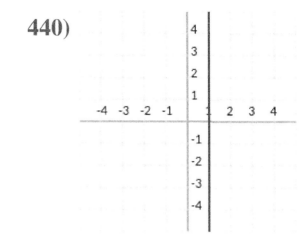

441)

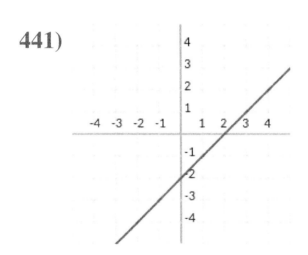

442)

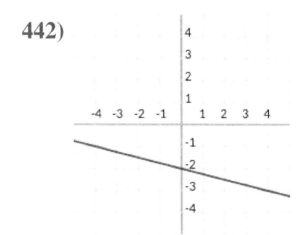

443)

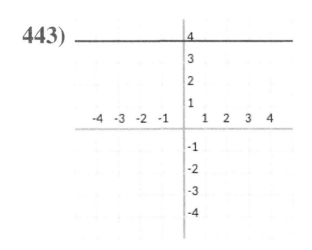

444)

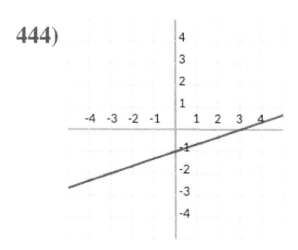

Linear Graphs

Give the equation for each line in the form y = m x + b

445)
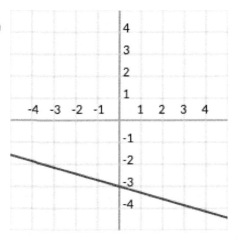

446)
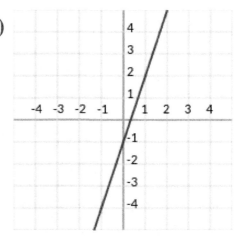

447)
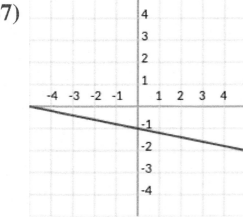

448)
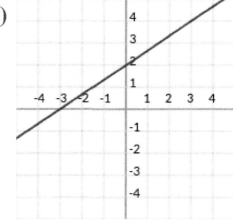

449)
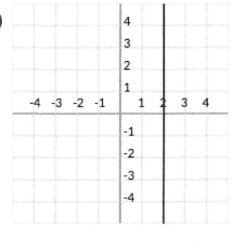

450)
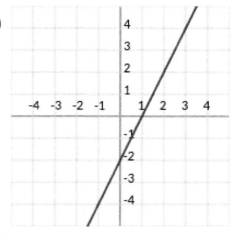

Linear Graphs

Give the equation for each line in the form y = m x + b

451)

452)

453)

454)

455)

456)

Linear Graphs

Give the equation for each line in the form y = m x + b

457)

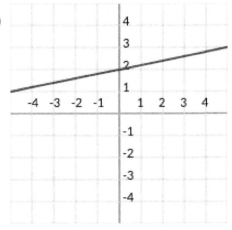

458)

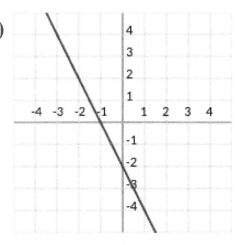

459)

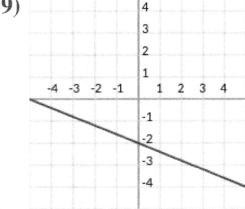

460)

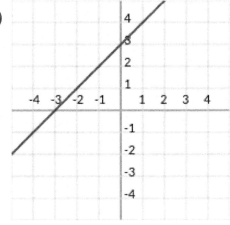

461)

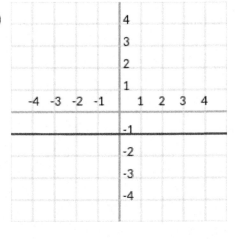

462)

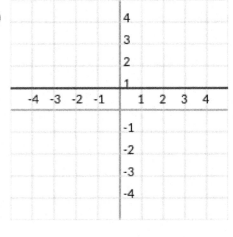

Linear Graphs

Give the equation for each line in the form y = m x + b

463)

464)

465)

466)

467)

468)

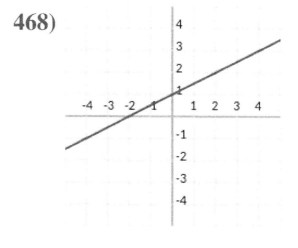

Linear Graphs

Give the equation for each line in the form y = m x + b

469)

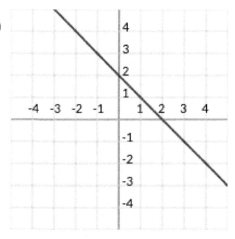

470)

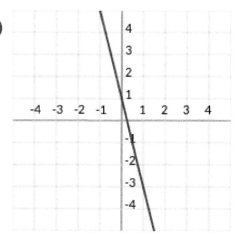

471)

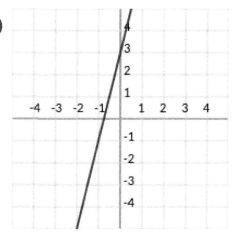

472)

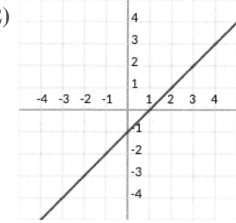

473)

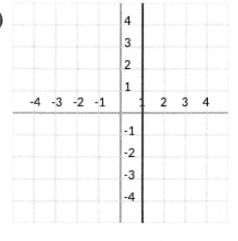

474)

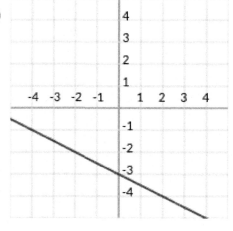

Linear Graphs

Give the equation for each line in the form y = m x + b

475)

476)

477)

478)

479)

480)

Linear Graphs

Give the equation for each line in the form y = m x + b

481)

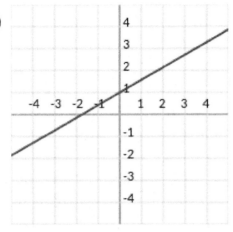

482)

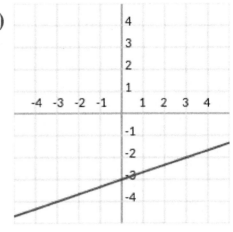

483)

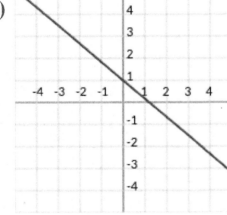

484)

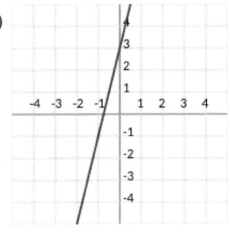

485)

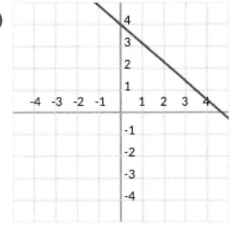

486)

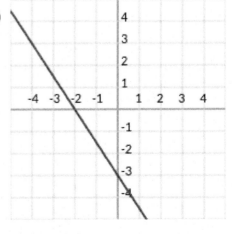

Linear Graphs

Give the equation for each line in the form y = m x + b

487)

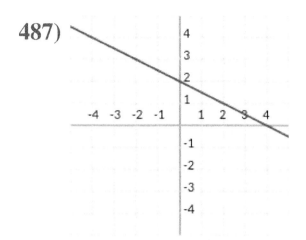

488)

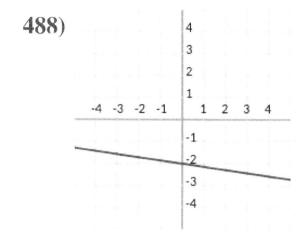

489)

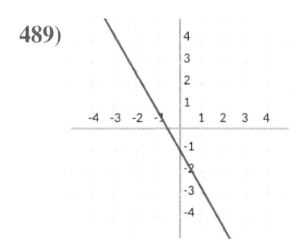

490)

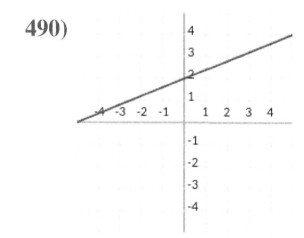

491)

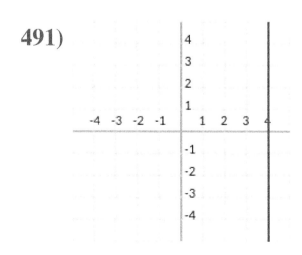

492)

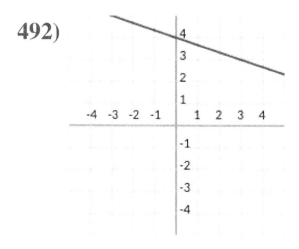

Linear Graphs

Give the equation for each line in the form y = m x + b

493)

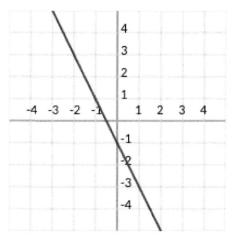

494)

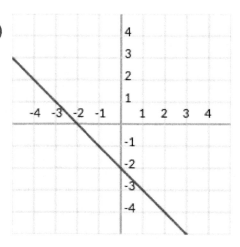

495)

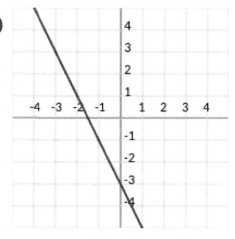

496)

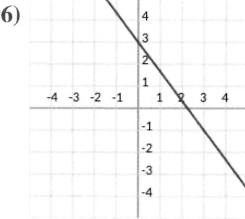

497)

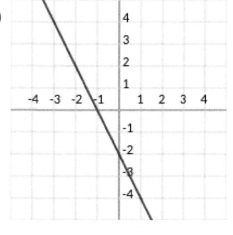

498)

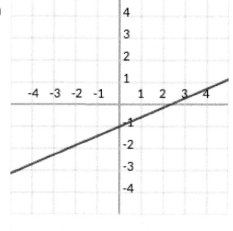

Give the equation for each line in the form y = m x + b

499)

500)

501)

502)

503)

504)

Linear Graphs

Give the equation for each line in the form y = m x + b

505)

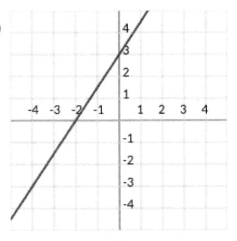

506)

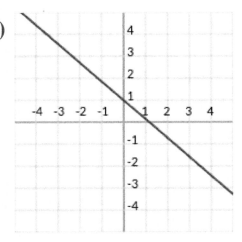

507)

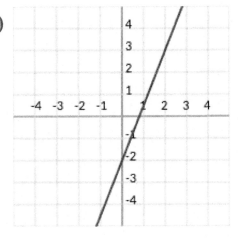

508)

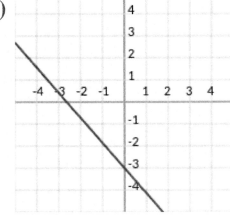

509)

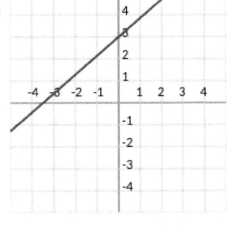

510)

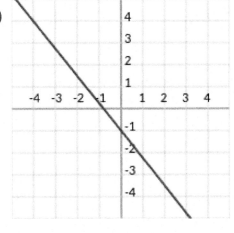

Linear Graphs

Give the equation for each line in the form y = m x + b

511)

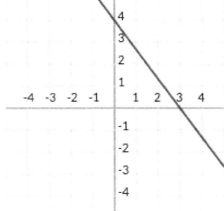

512)

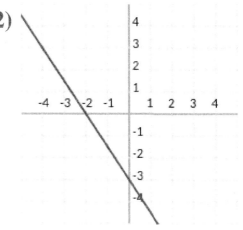

513)

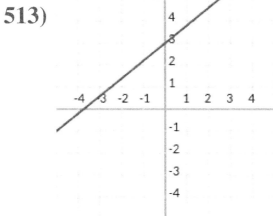

514)

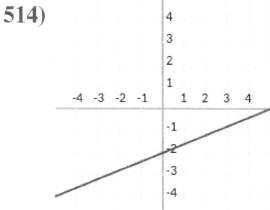

515)

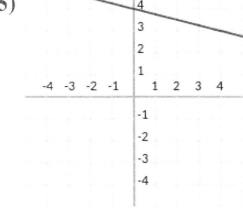

516)

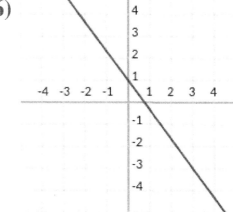

Linear Graphs

Give the equation for each line in the form y = m x + b

517)

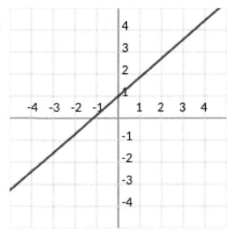

518)

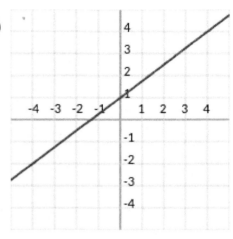

519)

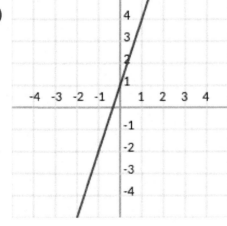

520)

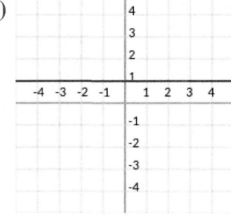

521)

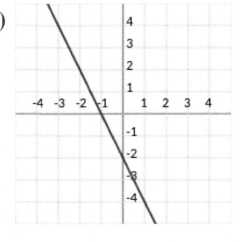

522)

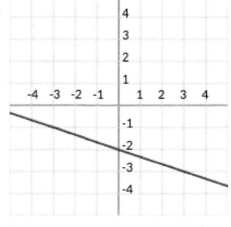

Linear Graphs

Give the equation for each line in the form y = m x + b

523)

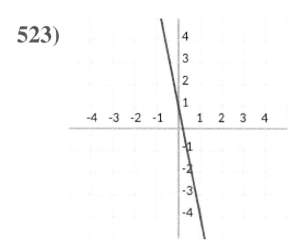

524)

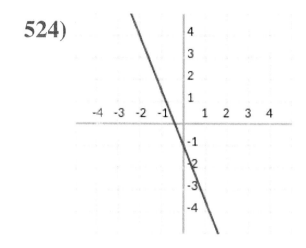

525)

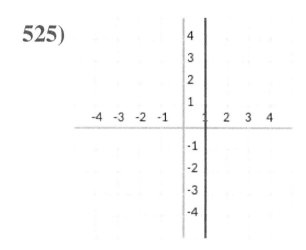

526)

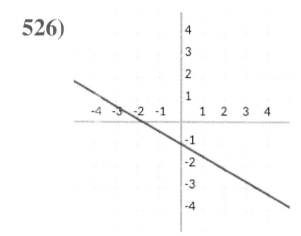

527)

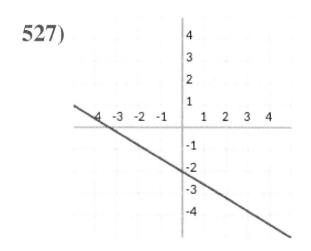

528)

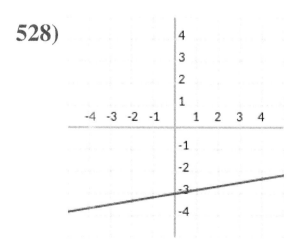

Linear Graphs

Give the equation for each line in the form y = m x + b

529)

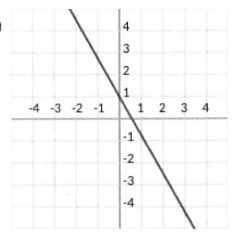

530)

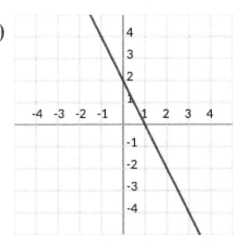

531)

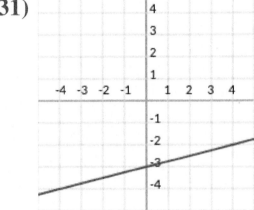

532)

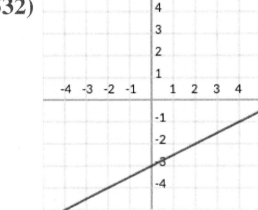

533)

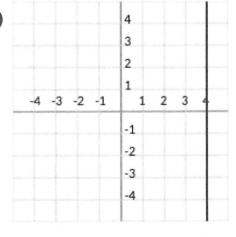

534)

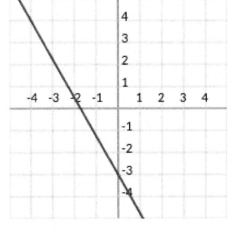

Bonus: Simple Linear Equations Sample

Enjoy this sample of *Algebra Book 1: Simple Linear Equations*

Solve for x

1) $-2x + 10 = 4$

2) $-15x + 16 = 16$

3) $-14x - 10 = 4$

4) $-4x + 12 = -13$

5) $x + 9 = -7$

6) $-13x + 11 = -13$

7) $13x + 4 = -12$

8) $-5x + 5 = -3$

9) $-3x - 1 = -2$

10) $3x + 12 = -14$

11) $-12x - 7 = -5$

12) $-9x - 2 = -10$

Algebra Book 1: Simple Linear Equations
80 puzzles in large print with solutions.

Available on Amazon Visit: smile.ws/ps9

Bonus: Hard Linear Equations Sample

Enjoy this sample of *Algebra Book 2: Hard Linear Equations*

Solve for x

1) $-4x + 6 = -8$

2) $10x - 4 = 2$

3) $-3x + 10 = 7$

4) $-15x - 7 = -8$

5) $-10x - 10 = -9$

6) $-6x + 9 = 7$

7) $4x + 4 = -15$

8) $-13x + 10 = -13$

9) $-9x - 5 = 7$

10) $-15x - 13 = -8$

11) $13x + 4 = 11$

12) $4x - 8 = 6$

Algebra Book 2: Hard Linear Equations
Over 1000 problems in large print with solutions.
Available on Amazon Visit: smile.ws/ pmx9

Bonus: Quadratic Factoring Sample

Enjoy this sample of *Algebra Book 3: Quadratic Factoring*

Factor these expressions

1) $x^2 + 6x + 8$

2) $x^2 + 2x - 15$

3) $x^2 + 9x + 20$

4) $x^2 - 5x + 6$

5) $x^2 - 3x - 4$

6) $x^2 - 6x + 5$

7) $x^2 + 5x + 4$

8) $x^2 - 5x + 6$

9) $x^2 - 5x + 6$

10) $x^2 + 4x + 3$

11) $x^2 - 9x + 20$

12) $x^2 + 8x + 12$

Answers for sample puzzles are in their respective full books.

Algebra Book 3: Quadratic Factoring
80 puzzles in large print with solutions.

Available on Amazon Visit: smile.ws/ps9

Bonus: Hard Sudoku Sample

How to play

A sudoku puzzle grid consists of nine rows and nine columns, arranged into nine sets of 3×3 inner-grids. The goal is to ensure each row, column, and inner-grid contain the numbers 1 to 9 with no repeats.

5				1			8	3
	2							
	6				3			
1			4	6		2		
							1	
	4			5				7
		7				5		8
		2		9	6			
	9	3		7				2

Bonus: Multi-Step Long Math Sample

A fun twist on math! Each problem has multiple steps.
Solve the first pair in each problem. Then take your answer and
use it with the next step in the same problem (and so on).

13. 2215
1605
+ 6828
_____

14. 4622
854
x   16
_____

15. 7555
5426
-   887
_____

22. 7262
4973
594
-   412
_____

23. 3264
4823
880
+ 2513
_____

24. 4869
943
62
x   9
_____

151. 6723
2415
3876
+ 4420
_____

152. 2821
206
9
x   9
_____

153. 7994
5354
1710
-   770
_____

Answers for sample puzzles are in their respective full books.

(Sample is non-sequential to reflect the variety in the book.)

1) $y = -\frac{1}{2}x - 2$ 2) $y = \frac{1}{5}x + 4$ 3) $y = x + 1$

4) $x = 4$ 5) $y = \frac{4}{3}x + 4$ 6) $y = \frac{4}{7}x + 2$

7) $y = -\frac{5}{7}x - 1$ 8) $y = 5x - 1$ 9) $y = \frac{6}{7}x + 1$

10) $y = \frac{2}{3}x + 1$ 11) $y = \frac{1}{3}x - 1$ 12) $y = \frac{1}{2}x + 4$

13) $y = \frac{2}{7}x - 3$ 14) $y = \frac{5}{3}x + 3$ 15) $y = 6x - 3$

16) $y = \frac{1}{3}x + 3$ 17) $y = -\frac{6}{5}x + 3$ 18) $y = -\frac{2}{3}x - 2$

19) $y = \frac{3}{2}x + 1$ 20) $y = 8x + 3$ 21) $y = \frac{7}{3}x + 1$

22) $y = -\frac{1}{3}x - 2$ 23) $y = \frac{7}{2}x + 2$ 24) $y = -x + 2$

25) $y = -\frac{1}{3}x + 4$ 26) $y = -x - 3$ 27) $x = 3$

28) $y = -\frac{1}{7}x + 4$ 29) $y = \frac{2}{3}x - 2$ 30) $y = \frac{5}{6}x + 3$

31) $y = \frac{6}{7}x + 2$ 32) $y = \frac{3}{2}x - 1$ 33) $y = 2x + 2$

34) $y = -\frac{1}{2}x - 2$ 35) $y = \frac{1}{5}x - 3$ 36) $y = \frac{5}{7}x + 2$

37) $y = 2x + 1$ 38) $x = 1$ 39) $y = -\frac{4}{3}x - 3$

40) $y = -\frac{2}{5}x + 4$ 41) $y = \frac{7}{3}x + 1$ 42) $y = x - 2$

43) $y = -7x + 4$ 44) $y = -\frac{6}{5}x - 1$ 45) $y = -\frac{8}{5}x + 1$

46) $y = x + 1$ 47) $y = \frac{1}{5}x + 1$ 48) $y = -\frac{1}{6}x + 1$

49) $x = -2$ 50) $y = -\frac{7}{6}x + 1$ 51) $y = \frac{2}{3}x + 4$

52) $y = -\frac{1}{5}x - 2$ 53) $y = x + 2$ 54) $x = 4$

55) $y = -\frac{5}{3}x - 3$ 56) $y = -\frac{1}{8}x - 3$ 57) $y = \frac{1}{7}x - 3$

58) $y = -\frac{8}{7}x - 1$ 59) $y = \frac{2}{7}x - 1$ 60) $y = \frac{3}{8}x + 4$

61) $y = -x - 3$ 62) $y = -3x - 2$ 63) $y = \frac{2}{7}x + 3$

64) $x = -2$ 65) $y = \frac{7}{2}x + 3$ 66) $y = \frac{5}{6}x + 1$

67) $x = -1$ 68) $y = 4x + 2$ 69) $x = 4$

70) $y = x + 2$ 71) $y = -x - 1$ 72) $y = \frac{1}{2}x + 1$

73) $y = \frac{3}{8}x + 2$ 74) $y = -\frac{2}{3}x - 2$ 75) $y = \frac{2}{3}x - 2$

76) $y = \frac{5}{2}x + 2$ 77) $y = \frac{1}{3}x + 3$ 78) $y = \frac{1}{2}x + 4$

79) $y = \frac{8}{7}x + 1$ 80) $y = \frac{7}{2}x + 3$ 81) $x = 3$

82) $y = -\frac{1}{4}x + 3$ 83) $y = \frac{2}{3}x + 4$ 84) $y = -x + 3$

85) $y = -\frac{8}{5}x + 1$ 86) $y = \frac{3}{4}x - 3$ 87) $y = \frac{3}{4}x + 3$

88) $y = \frac{6}{5}x - 3$ 89) $y = -\frac{1}{3}x + 4$ 90) $y = x + 1$

91) $y = -2x + 4$ 92) $x = 4$ 93) $y = -\frac{1}{7}x + 3$

94) $y = 3x + 4$ 95) $y = 2x + 4$ 96) $y = \frac{1}{2}x - 3$

97) $y = \frac{4}{5}x - 2$ 98) $y = -\frac{7}{3}x - 3$ 99) $y = \frac{6}{7}x - 3$

100) $y = -\frac{3}{7}x - 1$ 101) $y = -\frac{1}{3}x + 1$ 102) $y = -7x - 2$

103) $y = \frac{1}{4}x - 1$ 104) $y = -\frac{6}{7}x + 2$ 105) $y = -\frac{1}{5}x + 3$

106) $y = \frac{7}{3}x + 3$ 107) $y = -\frac{1}{2}x - 2$ 108) $y = -6x - 2$

109) $y = -\frac{5}{3}x - 2$ 110) $y = 7x + 2$ 111) $y = -\frac{1}{3}x + 3$

112) $y = -x + 4$ 113) $y = -\frac{1}{4}x + 2$ 114) $y = x - 2$

115) $y = -5x + 3$ 116) $y = -\frac{2}{5}x - 1$ 117) $y = \frac{4}{5}x - 2$

118) $x = -3$ 119) $y = -\frac{7}{5}x - 3$ 120) $y = -\frac{1}{3}x - 3$

121) $y = -\frac{1}{4}x + 4$ 122) $y = 7x - 3$ 123) $y = -\frac{4}{7}x + 1$

124) $y = 5x + 2$ 125) $y = \frac{1}{3}x + 1$ 126) $y = \frac{5}{6}x + 4$

127) $y = \frac{3}{7}x + 4$ 128) $y = \frac{4}{3}x + 1$ 129) $y = -2x + 2$

130) $y = \frac{4}{5}x + 3$ 131) $x = -3$ 132) $y = \frac{3}{4}x + 1$

133) $y = -\frac{3}{4}x + 1$ 134) $y = -\frac{3}{5}x + 1$ 135) $y = -\frac{7}{2}x - 2$

136) $y = x - 2$ 137) $y = x + 1$ 138) $y = x + 3$

139) $y = -4x + 2$ 140) $x = 2$ 141) $y = -x - 1$

142) $x = -1$ 143) $y = \frac{7}{8}x - 3$ 144) $y = \frac{4}{3}x - 2$

145) $y = \frac{5}{6}x + 1$ 146) $y = -\frac{2}{3}x - 1$ 147) $y = \frac{8}{7}x + 3$

148) $y = -2x - 3$ 149) $y = -\frac{5}{7}x + 2$ 150) $y = -\frac{2}{5}x + 3$

151) $y = \frac{1}{3}x + 1$ 152) $y = -\frac{7}{6}x + 3$ 153) $y = -x + 4$

154) $y = x + 2$ **155)** $y = -\frac{5}{7}x + 1$ **156)** $y = x - 2$

157) $y = -\frac{5}{6}x - 1$ **158)** $y = \frac{6}{5}x - 1$ **159)** $y = -\frac{2}{7}x + 3$

160) $y = 3x + 2$ **161)** $x = 2$ **162)** $y = -\frac{8}{7}x + 1$

163) $y = -\frac{8}{7}x + 4$ **164)** $y = -7x + 3$ **165)** $y = x + 2$

166) $y = 3x - 1$ **167)** $y = 2x + 2$ **168)** $y = x - 1$

169) $y = -4x + 3$ **170)** $y = -\frac{2}{7}x - 3$ **171)** $y = -\frac{3}{7}x + 1$

172) $y = -\frac{4}{5}x - 2$ **173)** $y = \frac{4}{3}x - 1$ **174)** $y = -4x + 1$

175) $y = \frac{6}{7}x + 2$ **176)** $y = -5x - 2$ **177)** $y = x + 1$

178) $x = -2$ **179)** $y = -6x - 1$ **180)** $y = -\frac{2}{7}x - 2$

181) $y = -\frac{7}{3}x + 4$ **182)** $y = -x - 1$ **183)** $y = \frac{7}{6}x + 4$

184) $y = -\frac{7}{5}x + 2$ **185)** $y = 4x + 1$ **186)** $y = x - 2$

187) $y = -\frac{3}{2}x + 1$ **188)** $y = -x - 2$ **189)** $y = -\frac{3}{4}x + 3$

190) $y = -4x - 3$ **191)** $y = \frac{5}{7}x + 3$ **192)** $y = x + 4$

193) $y = -\frac{5}{3}x + 4$ **194)** $y = \frac{8}{5}x - 1$ **195)** $y = -x + 3$

196) $y = x - 3$ **197)** $y = -x + 2$ **198)** $y = -x + 3$

199) $y = -\frac{1}{4}x + 2$ **200)** $y = \frac{1}{2}x - 3$ **201)** $y = \frac{5}{8}x + 2$

202) $y = \frac{2}{7}x + 3$ **203)** $y = -3x + 3$ **204)** $y = x - 1$

205) $y = -\frac{2}{3}x + 3$ **206)** $y = -6x - 1$ **207)** $y = \frac{1}{6}x + 4$

208) $y = \frac{2}{7}x + 1$ **209)** $y = -\frac{5}{7}x - 3$ **210)** $y = \frac{5}{3}x + 4$

211) $y = -\frac{2}{3}x + 2$ **212)** $y = x + 3$ **213)** $y = -\frac{5}{4}x + 4$

214) $y = -\frac{5}{2}x + 4$ **215)** $y = -\frac{6}{7}x - 1$ **216)** $y = -\frac{3}{2}x - 3$

217) $y = -\frac{6}{5}x + 4$ **218)** $y = -\frac{5}{3}x - 3$ **219)** $y = \frac{7}{3}x + 1$

220) $y = -\frac{7}{8}x + 1$ **221)** $y = -\frac{8}{3}x + 3$ **222)** $y = -\frac{6}{5}x + 3$

223) $y = x - 1$ **224)** $y = 2x + 2$ **225)** $y = \frac{3}{4}x + 2$

226) $y = -4x + 4$ **227)** $y = \frac{8}{3}x - 3$ **228)** $y = -\frac{7}{4}x + 1$

229) $y = \frac{5}{7}x - 1$ **230)** $y = \frac{2}{3}x - 3$ **231)** $y = x - 2$

232) $y = 3x - 2$ **233)** $y = -\frac{4}{5}x - 1$ **234)** $y = \frac{1}{2}x + 4$

235) $x = 1$ **236)** $y = \frac{6}{5}x - 1$ **237)** $x = -2$

238) $y = \frac{1}{4}x + 2$ **239)** $x = -1$ **240)** $y = -\frac{8}{7}x + 1$

241) $y = \frac{1}{3}x + 2$ **242)** $y = -\frac{3}{7}x - 3$ **243)** $y = \frac{5}{6}x + 2$

244) $y = \frac{1}{2}x - 3$ **245)** $y = \frac{2}{5}x - 3$ **246)** $y = -\frac{3}{2}x + 3$

247) $y = -x + 3$ **248)** $x = 4$ **249)** $y = \frac{6}{7}x + 4$

250) $y = -x - 1$ **251)** $x = -3$ **252)** $y = \frac{1}{6}x + 3$

253) $y = -6x - 1$ **254)** $y = -4x + 2$ **255)** $y = \frac{7}{8}x - 1$

256) $y = -\frac{7}{6}x + 1$ **257)** $y = \frac{2}{3}x - 3$ **258)** $y = -x - 3$

259) $y = 3x + 1$ **260)** $y = x + 1$ **261)** $y = -\frac{7}{6}x + 1$

262) $y = \frac{3}{8}x + 1$ **263)** $y = -\frac{5}{7}x + 1$ **264)** $y = -\frac{5}{2}x - 1$

265) $y = \frac{5}{3}x - 3$ **266)** $y = 2x + 1$ **267)** $x = -1$

268) $y = \frac{3}{5}x - 2$ **269)** $y = -\frac{1}{5}x + 3$ **270)** $y = -\frac{2}{7}x - 3$

271) $y = -\frac{3}{4}x + 4$ **272)** $y = 2x - 1$ **273)** $x = -3$

274) $y = \frac{8}{3}x + 1$ **275)** $y = -\frac{7}{2}x - 2$ **276)** $y = -\frac{1}{4}x + 4$

277) $y = -\frac{3}{5}x - 2$ **278)** $y = \frac{5}{2}x + 1$ **279)** $y = \frac{5}{3}x + 2$

280) $x = -1$ **281)** $y = -x + 1$ **282)** $y = -\frac{1}{3}x + 3$

283) $y = -2x + 4$ **284)** $y = \frac{7}{2}x + 2$ **285)** $x = 3$

286) $y = -x + 3$ **287)** $y = \frac{4}{7}x - 2$ **288)** $y = \frac{1}{2}x - 1$

289) $y = -8x - 3$ **290)** $y = x - 3$ **291)** $y = -5x - 2$

292) $y = \frac{3}{7}x + 4$ **293)** $y = \frac{1}{7}x - 3$ **294)** $y = x - 3$

295) $x = 2$ **296)** $y = -\frac{4}{7}x + 2$ **297)** $y = -\frac{2}{3}x - 1$

298) $y = \frac{1}{2}x + 1$ **299)** $y = -\frac{5}{3}x + 3$ **300)** $y = -2x - 2$

301) $y = \frac{3}{2}x - 3$ **302)** $y = \frac{1}{7}x + 3$ **303)** $x = 4$

304) $y = x - 1$ **305)** $y = -\frac{7}{8}x + 4$ **306)** $y = \frac{3}{2}x - 1$

307) $y = \frac{4}{3}x - 1$ **308)** $y = -\frac{5}{6}x - 2$ **309)** $y = \frac{4}{7}x - 1$

310) $y = \frac{1}{2}x + 1$ **311)** $y = \frac{2}{3}x + 4$ **312)** $y = \frac{3}{2}x - 2$

313) $y = -\frac{5}{2}x + 4$ **314)** $y = -3x + 3$ **315)** $y = -6x + 1$

316) $y = 2x - 3$ **317)** $y = -\frac{5}{2}x - 3$ **318)** $x = -2$

319) $y = -x + 2$ **320)** $y = -\frac{3}{5}x + 1$ **321)** $y = -3x + 3$

322) $y = -x - 2$ **323)** $y = 4x + 2$ **324)** $y = x + 1$

325) $y = \frac{1}{4}x - 3$ **326)** $y = x + 4$ **327)** $y = \frac{5}{4}x - 3$

328) $y = -\frac{1}{6}x - 1$ **329)** $y = \frac{4}{5}x + 3$ **330)** $y = 6x - 2$

331) $y = -x + 1$ **332)** $y = \frac{1}{2}x - 3$ **333)** $y = 8x + 1$

334) $y = -\frac{8}{7}x + 3$ **335)** $y = \frac{3}{4}x - 1$ **336)** $x = -1$

337) $y = -\frac{3}{2}x + 4$ **338)** $y = -x + 1$ **339)** $y = 5x + 3$

340) $y = -\frac{2}{7}x + 3$ **341)** $y = \frac{7}{6}x + 4$ **342)** $y = \frac{3}{2}x - 1$

343) $y = x - 3$ **344)** $y = \frac{1}{6}x + 1$ **345)** $y = \frac{7}{2}x + 4$

346) $y = 4x + 4$ **347)** $y = \frac{3}{7}x - 2$ **348)** $y = \frac{6}{7}x + 1$

349) $y = \frac{3}{7}x + 3$ **350)** $y = -\frac{3}{7}x + 3$ **351)** $y = -\frac{3}{4}x + 2$

352) $y = -\frac{4}{5}x + 1$ **353)** $y = \frac{1}{3}x - 2$ **354)** $y = \frac{1}{2}x - 3$

355) $y = -\frac{1}{2}x + 4$ **356)** $y = x + 3$ **357)** $y = \frac{7}{4}x - 1$

358) $y = -\frac{7}{5}x - 1$ 359) $y = -3x + 3$ 360) $y = -\frac{1}{5}x - 1$

361) $y = \frac{5}{6}x - 2$ 362) $y = \frac{7}{3}x + 2$ 363) $y = x + 2$

364) $y = -\frac{1}{2}x - 3$ 365) $y = -\frac{1}{2}x - 2$ 366) $y = \frac{3}{7}x - 1$

367) $y = -2x + 2$ 368) $y = -\frac{3}{5}x + 3$ 369) $y = x + 2$

370) $y = \frac{8}{5}x + 1$ 371) $x = 2$ 372) $y = -5x - 3$

373) $y = -\frac{7}{4}x + 4$ 374) $y = -\frac{7}{3}x + 4$ 375) $y = -\frac{7}{8}x - 1$

376) $y = x + 2$ 377) $y = x + 1$ 378) $y = -\frac{5}{4}x + 1$

379) $y = \frac{4}{5}x + 4$ 380) $x = 2$ 381) $y = 5x + 4$

382) $y = -\frac{5}{3}x + 2$ 383) $y = \frac{2}{3}x + 2$ 384) $y = \frac{5}{7}x + 4$

385) $y = -x - 1$ 386) $y = 2x + 2$ 387) $y = \frac{1}{4}x + 2$

388) $y = -\frac{7}{3}x + 4$ 389) $y = \frac{1}{5}x + 3$ 390) $y = 2x - 2$

391) $y = \frac{7}{2}x - 1$ 392) $y = \frac{1}{4}x - 3$ 393) $x = -2$

394) $y = -3x + 2$ 395) $y = \frac{8}{3}x + 2$ 396) $y = -\frac{8}{7}x + 1$

397) $y = x + 4$ 398) $y = \frac{5}{6}x + 3$ 399) $y = -6x - 2$

400) $x = 2$ 401) $y = 6x - 3$ 402) $y = \frac{3}{2}x + 4$

403) $y = \frac{2}{3}x - 2$ 404) $y = -\frac{1}{6}x - 2$ 405) $x = -3$

406) $y = \frac{2}{3}x + 4$ 407) $y = -x + 3$ 408) $y = 5x + 2$

409) $x = 2$

410) $y = -\frac{7}{2}x + 2$

411) $y = -\frac{1}{7}x + 4$

412) $x = -3$

413) $y = -\frac{1}{3}x + 3$

414) $y = -\frac{4}{7}x + 1$

415) $y = -\frac{1}{3}x + 3$

416) $y = -x - 1$

417) $y = -\frac{1}{2}x + 3$

418) $y = -x + 4$

419) $y = -\frac{3}{5}x - 2$

420) $x = -2$

421) $y = -\frac{1}{7}x - 2$

422) $y = -\frac{4}{7}x + 2$

423) $y = \frac{1}{2}x + 3$

424) $y = \frac{1}{3}x + 4$

425) $y = -x - 1$

426) $y = -\frac{1}{4}x + 4$

427) $y = -\frac{2}{7}x - 1$

428) $y = 3x + 4$

429) $x = -1$

430) $y = -\frac{1}{3}x - 3$

431) $y = \frac{8}{5}x - 1$

432) $y = -\frac{3}{8}x - 1$

433) $y = \frac{5}{6}x + 3$

434) $y = -2x + 2$

435) $y = -6x + 2$

436) $y = -2x + 1$

437) $y = \frac{2}{3}x - 2$

438) $y = -\frac{2}{3}x - 1$

439) $y = \frac{1}{7}x + 1$

440) $x = 1$

441) $y = x - 2$

442) $y = -\frac{1}{4}x - 2$

443) $y = x + 4$

444) $y = \frac{1}{3}x - 1$

445) $y = -\frac{2}{7}x - 3$

446) $y = 3x - 1$

447) $y = -\frac{1}{5}x - 1$

448) $y = \frac{2}{3}x + 2$

449) $x = 2$

450) $y = 2x - 2$

451) $y = 7x + 1$

452) $y = \frac{1}{6}x - 3$

453) $y = \frac{1}{2}x + 3$

454) $y = \frac{5}{3}x + 2$

455) $y = 2x + 2$

456) $y = -x - 1$

457) $y = \frac{1}{5}x + 2$

458) $y = -2x - 2$

459) $y = -\frac{2}{5}x - 2$

460) $y = x + 3$ **461)** $y = x - 1$ **462)** $y = x + 1$

463) $y = \frac{3}{7}x + 4$ **464)** $x = -3$ **465)** $y = -x + 1$

466) $y = -\frac{2}{3}x + 4$ **467)** $y = \frac{5}{6}x - 2$ **468)** $y = \frac{1}{2}x + 1$

469) $y = -x + 2$ **470)** $y = -4x + 1$ **471)** $y = 4x + 3$

472) $y = x - 1$ **473)** $x = 1$ **474)** $y = -\frac{1}{2}x - 3$

475) $y = -\frac{1}{2}x + 4$ **476)** $y = -\frac{7}{3}x - 3$ **477)** $y = \frac{1}{6}x - 2$

478) $y = \frac{5}{7}x - 1$ **479)** $y = \frac{6}{5}x - 3$ **480)** $y = -\frac{3}{5}x + 1$

481) $y = \frac{4}{7}x + 1$ **482)** $y = \frac{1}{3}x - 3$ **483)** $y = -\frac{5}{6}x + 1$

484) $y = 4x + 3$ **485)** $y = -\frac{6}{7}x + 4$ **486)** $y = -\frac{3}{2}x - 3$

487) $y = -\frac{1}{2}x + 2$ **488)** $y = -\frac{1}{7}x - 2$ **489)** $y = -\frac{7}{4}x - 1$

490) $y = \frac{2}{5}x + 2$ **491)** $x = 4$ **492)** $y = -\frac{1}{3}x + 4$

493) $y = -2x - 1$ **494)** $y = -x - 2$ **495)** $y = -2x - 3$

496) $y = -\frac{4}{3}x + 3$ **497)** $y = -2x - 2$ **498)** $y = \frac{3}{7}x - 1$

499) $x = 3$ **500)** $y = -\frac{6}{7}x + 2$ **501)** $y = -2x - 2$

502) $y = \frac{3}{4}x - 2$ **503)** $y = -\frac{5}{7}x - 2$ **504)** $y = \frac{1}{7}x + 2$

505) $y = \frac{3}{2}x + 3$ **506)** $y = -\frac{6}{7}x + 1$ **507)** $y = \frac{5}{2}x - 2$

508) $y = -\frac{8}{7}x - 3$ **509)** $y = \frac{6}{7}x + 3$ **510)** $y = -\frac{5}{4}x - 1$

Answers

511) $y = -\frac{4}{3}x + 4$ 512) $y = -\frac{3}{2}x - 3$ 513) $y = \frac{4}{5}x + 3$

514) $y = \frac{2}{5}x - 2$ 515) $y = -\frac{1}{4}x + 4$ 516) $y = -\frac{4}{3}x + 1$

517) $y = \frac{6}{7}x + 1$ 518) $y = \frac{3}{4}x + 1$ 519) $y = 3x + 1$

520) $y = x + 1$ 521) $y = -2x - 2$ 522) $y = -\frac{1}{3}x - 2$

523) $y = -5x + 1$ 524) $y = -\frac{5}{2}x - 1$ 525) $= 1$

526) $y = -\frac{4}{7}x - 1$ 527) $y = -\frac{3}{5}x - 2$ 528) $y = \frac{1}{6}x - 3$

529) $y = -\frac{7}{4}x + 1$ 530) $y = -2x + 2$ 531) $y = \frac{1}{4}x - 3$

532) $y = \frac{1}{2}x - 3$ 533) $x = 4$ 534) $y = -\frac{7}{4}x - 3$

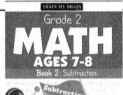

Printed in Great Britain
by Amazon

35593723R00064